THE NEW ARCADIA:
TAHITI'S CURSED MYTH

MONIQUE LAYTON

 **FriesenPress**

Suite 300 - 990 Fort St
Victoria, BC, Canada, V8V 3K2
www.friesenpress.com

Copyright © 2015 by Monique Layton
First Edition — 2015

Other books by the author:
Street Women and the Art of Bullshitting (2010)
Notes from Elsewhere. Travel and Other Matters (2011)
Translation: Claude Lévi-Strauss. *Structural Anthropology 2* (1976)

All rights reserved.

No part of this publication may be reproduced in any form, or by any means, electronic or mechanical, including photocopying, recording, or any information browsing, storage, or retrieval system, without permission in writing from the publisher.

ISBN
978-1-4602-6859-9 (Hardcover)
978-1-4602-6860-5 (Paperback)
978-1-4602-6861-2 (eBook)

1. History, Expeditions & Discoveries

Distributed to the trade by The Ingram Book Company

TABLE OF CONTENTS

Reviews for Notes from Elsewhere.
Travel and Other Matters..vii

Acknowledgements:...ix

A Personal Introduction to the Tahitian Myth 1

Part One: THE MYTH-MAKERS 9

[I] The Long Way to Polynesia 11
    Scurvy...13
    Terra Australis Incognita 18
    The New Explorers..22
    Polynesian Seafarers 29

[II] Otaheite, Earthly Paradise................................39
    A Tropical Arcadia...41
    The Noble Savage 53
    Going Native...59

[III] The Vahine 63
    Tahiti Through the Silver Screen........................64
    In the Eye of the Beholder 69
    The "Wifies"...75
    The Vahine and Tourism 84
    Reversal of Desires 85

[IV] Cythera Revisited. 91
    Human Sacrifice, Cannibalism, and Infanticide 91
    "The Allurements of Dissipation". 96
    Skirmishes 103
    Thievery 107

Part Two: RUINATION 113

[V] The Warnings 117

[VI] The Crucifix11 125
    The Missions' Impact 126
    The London Missionary Society 130
    The Catholic Picpusians 135
    The Case for the Missionaries 138

[VII] The Diseases 147
    The Pox 148
    Epidemics 153
    Population Count 156

[VIII] And Other Woes 161
    The Itinerants: Traders, Whalers, and Others 161
    Slavery 163
    Changing Technology 165
    The Ignoble Savage 168

Part Three: FROM OTAHEITE TO THE NEW TAHITI .. 171

[IX] Resistance, Nuclear Testing, and Self-Determination .. 173
    Distress and Rebellion 175
    Colonial Days 179
    Nuclear Testing 182
    Self-Determination and the Indépendantiste Movement 185

[X] Language, Literature, and Politics 191
    Reo Ma'ohi 194
    The Oral Tradition 196

Tahitian Literary Forms 200
The Literature of Resentment 202

[XI] Shattered Dreams and Materena's World 205
Spitz and Vaite 205
History and Politics 210
Connection with the Land 212
Love and Attachment 213
The Demis 214
The Vahine and her Life 217

[XII] Tourism and Noa Noa 227
Competin' for the Yankee Dollar 232
Kite Makers vs. Directors of Tourism Development 241
Traditions 246
Kite-flying 246
Canoes 247
Tattoos 252
Dances 257
Food 260

[XIII] The Quandary 269

Notes 281

Glossary 295

Bibliography 297

REVIEWS FOR NOTES FROM ELSEWHERE. TRAVEL AND OTHER MATTERS.

"This wide-ranging book—literally and metaphorically—from Morocco to Bangkok, from theories of travel and cultural engagements to the fragile details of elusive memory reveals a singular voice and a capacious mind."

Eleanor Wachtel (Author of Writers and Company, Original Mind, and Random Illuminations).

"Common to all travels, journeys, voyages, and sojourns is the search for the familiar. This search is Monique Layton's major contribution. Elsewhere, for her, is caught in a web of memory, a rich mélange of philosophical musings and literary allusions. Her text is part ethnography and part gentle mockery as if surprised by her own presence in her varying realities. All this adds up to considerable reading pleasure."

Elvi Whittaker (Author of A Baltic Odyssey, The Mainland Haole, and The Silent Dialogue).

"Layton constantly offers rich, sensory details of her journeys… [Her] observations are a candid, smartly edited mix of positive and negative details. [Her] blend of historical, literary and political references throughout helps put her keen personal observations in context… The memoir offers striking details about less-traveled locations and thought-provoking commentary on the difficulties of understanding a culture other than one's own."

Kirkus Review

ACKNOWLEDGEMENTS:

I HAVE, AS USUAL, RELIED on the generosity of my friends. Elvi Whittaker, with her Polynesian field experience and her careful guidance since my student days; Lee Southern, with his terrific eye for inconsistencies; and Patricia Fung, with her knowledge of island paradises have my gratitude for accepting to read my first draft. I should also mention Miriam Kahn (University of Washington) and Masako Ikeda (University of Hawai'i Press) who kindly provided comments and encouragements.

It has been my good fortune to spend the last three years in excellent company. First, Bligh, Banks, Bougainville, Commerson, and many others of lesser rank, with all their faults and their extraordinary bravery, and most of all, the admirable James Cook. Next, two scholars who have particularly informed my writing: Professors Anne Salmond on eighteenth-century navigators and Miriam Kahn on contemporary Tahiti. Finally, Tahitian writers, struggling to make sense of their colonial past, adjusting to ensure the passage from an oral tradition to the written word, resenting having to use a foreign language to express an essentially Ma'ohi vision, and often refusing to compromise while defining their aspirations.

This book is for John.
Vancouver, 2015

A PERSONAL INTRODUCTION TO THE TAHITIAN MYTH

IN 2011, I SUFFERED MULTIPLE fractures in an accident. For the five months I spent in a number of hospitals, I wrote about injuries, sickness, recovery, medical and nursing staff, as well as old age and death, attempting to keep an open mind and the necessary distance to write an ethnographic account of hospital life. Until, one day, my mind came up with an antipodal image: Tahiti—of which I knew nothing beyond a vague South Seas cliché.

During the nine months of rehabilitation that followed my discharge from hospital, I read about Tahiti and her islands. As a long-time insomniac, most of my nights were also spent reading. I started with Pierre Loti and Bengt Danielsson, and ended up with Anne Salmond and Miriam Kahn. And, of course, I read everything else I could lay my hands on. My readings were not encyclopedic, but they were enough to help me define an outline for the evolution of the Tahitian myth by which so many have been seduced. Mostly, I pondered the apparent contradiction of a reputation as an earthly paradise leading to so much suffering and heartache for its inhabitants.

Finally, I went twice to Tahiti, in 2012 and 2014. Still using a walker and suffering from peripatetic pains, I did and saw very little of what I would so dearly have liked. But, each time our ship entered a bay and reached a port, I was immensely moved, thinking that perhaps *there,* in that very place where I now found myself, they had faced one another, the navigators in their three-masted barques and the islanders in their outrigger canoes, neither knowing what to expect from the other. It happened everywhere we went, in Vaitape (Bora Bora), in Opunohu Bay (Moorea), in Uturoa

(Raiatea), in Papeete (Tahiti), in Fakarava (Tuamotu), and in Taiohae Bay (Nuku Hiva, in the Marquesas). It was a fiction, of course, because the first visitors did not actually land in those very spots; but they could have.

I have visited other islands with their own claim to beauty. Indeed, when Christopher Columbus first caught sight of Cuba in 1492, he is supposed to have said that it was "the most beautiful land ever seen." I can confirm that it is beautiful and possesses perhaps the whitest and finest sand in the world, as well as some of the most derelict buildings still attesting to a former grandeur.

Although lacking the culture and history of Tahiti, the volcanic and coral archipelago of the Seychelles is just as beautiful, with its enormous ancient granite rocks emerging from bright aqua waters, its bird islands, and its Vallée de Mai National Park. The delightful island of Jersey, even if a little too civilized and proper, also has its attractions. Another, Vancouver Island, also reputed for its charms, has the singularity of being home to our youngest son and his family, and as such holds a special place in our hearts. As well, I have been a visitor to several of the Hawaiian Islands over time, and do not begrudge them my admiration. This is to say that I am not a newcomer to the specific attraction of islands. However, nowhere have I felt such emotion when nearing their shores as I did in the Society Islands and the Marquesas.

I admired the people's looks, natural good manners, and pleasant appearance. Their speech charmed me as a native French speaker. The enchanting rolling of the *r,* in which some see a deliberate and rebellious effort to "tahitianize" the French language, made it all the more delightful to my ears. When alone, I even practiced the pronunciation. In other words, I fell in love with Tahiti, her islands, and her people.

There was perhaps an absurdly sentimental element in my feelings, one I might partially attribute to my own upheaval during the preceding period. I cannot deny that it tainted my readings and writings about a land that, in theory, has nothing to do with my own history. But I felt intimately connected to something I could not name, because falling in love is the most irrational occurrence to which one can aspire.

What has not already been written about that part of the world? Very little, I should think. The navigators themselves, whose vessels succeeded one other at an extraordinary rate in Tahiti, wrote detailed reports at the end of the eighteenth century. Some twenty ships from three European nations anchored in Tahiti within a quarter of a century. For the times and in that part of the world, it must have felt like Grand Central Station.

At anchor in the Society Islands. (Sketch by John Cleveley)

All these ships carried on board men who charted the island's shores, observed its natural features, described the customs and events, discovered and classified new species, and wrote abundantly about their encounters with the natives. Charmed by what they found there, some called the island New Cythera in reference with its connection to Venus, while others referred to it as Paradise on Earth, More's Utopia, the New Arcadia, and so on. William Hodges and John Webber's paintings only served to feed the Europeans' imagination and anchor their vision, often bathed in then familiar classical references.

The two Tahitians who were taken to Europe, Ahutoru to Paris by Bougainville in 1768 and Mai or Omai to London by Captain Furneaux in 1774, charmed everyone with their countenance, fitting in with the image of the Noble Savage, a popular construct of the time. By then, the fashion was to admire Tahiti and long for her simple natural life.

Upon reading Bougainville's published journal, Diderot, the French *Encyclopédiste,* wrote it was the first time his readings had tempted him to visit a country other than his own. In England, Dr. Johnson was equally eager to visit and study Tahiti (or New Zealand); it is only James Boswell's reluctance to engage in such a journey that, unfortunately, put an end to Dr. Johnson's plans and our likely pleasure in reading about their adventures and discoveries.

Since those early days, writers and artists have taken upon themselves to continue spreading the idyllic image of Tahiti and the South Seas, particularly Loti, Gauguin, Melville, Stevenson, and Michener, among so many others. Later, historians and anthropologists brought back to public attention the navigators' voyages and their discoveries, reviving once more the interest in these islands. Others have examined specific aspects of the history and culture of Tahiti, and I have been particularly guided by Professor Salmond's profound knowledge of what governed the eighteenth century navigators' progress, and by Professor Kahn's study of modern Tahitians' relationship with their particular place.

From the start, I envisaged three parts to my own study. "The Myth-Makers" looks at the elaboration of the Tahitian myth in the European imagination. How was such an image of island perfection created? What choice of words caused Europeans who heard or read them to build in their minds the picture of a land full of beauty and innocence, an island of plenty where people lived freely and in harmony with nature? Like all myths, the evocation of Tahiti contains its own truth and is based on a natural and cultural reality; yet, as in all myths, this reality is intermingled with the smoke and mirrors of imagination and transcends their original explanatory purpose of justifying the present.

For this part, I relied on the extensive studies available, based on factual reports (voyage journals, letters, personal diaries) from captains, sailors, naturalists, ship surgeons, and others sailing on the *Dolphin,* the *Etoile,* the *Boudeuse,* the *Swallow,* the *Endeavour,* the *Resolution,* the *Discovery,* the *Bounty,* the *Providence,* the *Pandora,* and the *Lady Penhryn,* not to mention the Spanish *Aguilá* and *Jupiter.*

Inspired by Bougainville's spontaneous recognition of a newly found island in the vastness of the Pacific as *la Nouvelle Cythère,* I describe the nearly unanimous enchantment felt by those who first landed in Tahiti. After months at sea in harsh and often perilous conditions, they discovered a generous environment where lived handsome and hospitable people. Most of all, they found there an extraordinary culture that apparently permitted the women to give themselves with abandon to these unknown sailors. Nowhere else in their multiple voyages to faraway lands had these men met with such attractive women willing to engage in sexual encounters with so much apparent insouciance. They believed they had found the island of love.

Seduced by their own nostalgic desires, they also believed that they had stumbled upon Paradise before the fall of man. It sometimes took the briefest stay on the island to confirm them in their opinion. They based the latter on an ethnocentric notion of what a perfect society could be, unaware of—or perhaps unwilling to acknowledge—the evidence that Tahiti was not merely a transposition into the Pacific of the idealized European or Greek version of this perfect society they held as a model. They related what most fitted the image they had formed, and then, for good measure, they embellished their reports. What they also did, and some of the better and most clear-sighted men among them could already guess at the future, was put an impossible burden on Tahiti and her islands and do them a dreadful disservice.

The version of Tahiti that reached Europe was that of heaven on earth, in spite of a few passing reservations emerging in writings and comments, for it is hard to overlook entirely tribal warfare, human sacrifices, cannibalism, and infanticide, even if the culture somehow makes sense of them, as cultures are wont to do with all their manifestations. It almost seemed as if Europeans had an urgent need to depict only what fitted the myth they were eager to unleash upon their return home. The cooler heads, perhaps less educated in the classics, contributed their own version, merely that of a beautiful island, peopled with handsome and easy-going men and women living in harmony with nature. However, the version more universally adopted was the myth we know, an idealized version of the human condition and the impossible state of innocence imposed upon these islands. While most Europeans evoked a vague and emotional nostalgia for the lost innocence of mankind, others could already see that Tahitians had lost their own innocence—such as it was—when the first Englishman traded the first iron nail for sex.

None of us can know the ambivalent reality of these first encounters because we mostly ignore the undated oral epics of the Polynesians on the subject, and only rely on the written European version. The early contacts between the islanders and the white men have certainly been described in the most precise details, with many of the navigators trained in empirical observation and proving themselves to be excellent ethnographers, in spite of the difficulties created by the language. They also built theories on the Tahitians' culture and religion, and their artists' sketches depicted for us the life of the islands. Later, the first missionaries often provided useful information on the customs of their recent converts. Such are the sources we are still using to feed the myth, further enriched by fictional narratives about the South Seas.

The word having spread, Tahitians could only wait for the next sets of visitors. The title that imposed itself to me for this next part was "Ruination," a term imbued with an almost apocalyptic connotation. In it, I retrace the island's post-contact history: the spiritual and cultural losses following the arrival of the missionaries; the horrendous diseases and ensuing catastrophic population declines; the coming of the whalers, traders, settlers, and slavers; and the general demoralization.

In the third and last part, "From Otaheite to the New Tahiti," I review the reality of today's Tahiti and consider the historical, political, and intellectual evolutions from the days of the Europeans' first contact with Otaheite. These include momentous and extremely rapid changes, including the introduction of modern technology, the effects of French colonization and nuclear testing of later years, the passage from a strictly oral epic tradition to an insurgent written literature, the aggressive promotion of tourism as a revival of cultural expression, and, for some, the desire for independence. Unanimously, Tahitians call themselves *Ma'ohi,* and the language they speak and write is *reo ma'ohi,* their bond with their culture, their Polynesian ancestors, their land, and their everyday reality. In some cases, it also defines their politics. In this section, I review works by some Tahitian novelists to illustrate nationalistic trends.

For Europeans, Tahiti was the metaphor for a world deemed to be both ideal and primitive; a life in harmony with nature and a society at once hedonistic and amoral. They apprehended this duality as a symbol of Arcadian life in spite of a few systemic aberrations. Their terms of reference being their own societies, Europeans could only see Tahiti according to their own lights and through their artistic renditions. It is not surprising that Tahiti should be perceived in idealized forms.

Today, the myth—having lost its initial virtue of awakening Europeans to other possibilities than the ones they already knew—is more alive than ever, having fallen victim to the needs of tourism, the modern economic pillar of the islands. What earlier literature had limned as romantic exoticism has merely become a tantalizing promotion of holidays full of sea, sun, and sex.

Perhaps it is the same myth, adjusted to the taste of our times, but it no longer applies to the same people. Europeans saw this myth as the representation of an inherent part of Tahiti; Tahitians can only look with some derision and sometimes resentment upon such a foreign construct, and consider what it has cost them. At the same time, they also feed into the myth, fully aware that their economic welfare depends partly on the revenues of tourism.

For others, salvation from their current economic and political dependence on France could mean a return to the same Arcadian myth in their own lives. Their desire for a more autonomous status is matched by a return to a harmonious and self-sustaining life in accordance with nature, the one so much admired by the eighteenth century navigators.

This combination of topics, based on two and a half centuries of reports, studies, articles, essays, poetry, and novels written by Europeans, Americans, and Tahitians, is presented here as a continuous story. As the years pass and the evolution continues, more can be encompassed, particularly for a society still struggling to fit within its current metamorphosis, and sometimes feeling trapped into its travel poster image as a "paradise island." I tried to combine these many facets of the Tahitian culture into one study and, through it, trace, review, and measure the islanders' voyage across the white men's tempestuous interference.

PART ONE
THE MYTH-MAKERS

"An earthly paradise where men and women live happily in innocence, away from the corruption of civilization."

Comte de Bougainville. *Voyage around the World* (1772)

[1]
THE LONG WAY TO POLYNESIA

SOUTH OF THE EQUATOR, LOOKING lost in the vastness of the Pacific Ocean, sits a grouping of islands known today as French Polynesia. Seventy-three of these islands are inhabited—if, for a handful of them, fewer than ten people can be said to constitute a population. They are grouped into five archipelagoes, with different characteristics and different histories, several of which will be mentioned here with their main islands: the Society Islands, (Tahiti, Raiatea, Moorea, Bora Bora, Huahine), the Austral Islands (Rapa), the Tuamotu Islands (Rangiroa), the Gambier Islands (Mangareva), and the Marquesas Islands (Nuku Hiva, Hiva Oa).

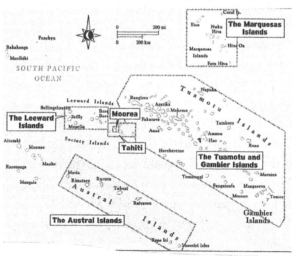

Map of French Polynesia

In 1768, when the French circumnavigator Bougainville arrived in Otaheite, as Tahiti was then known, he fell under the spell of what he called an "earthly paradise" and a "new Cythera," the Greek island reputed to be the birthplace of Venus. Like several of his contemporaries, the English navigators and naturalists who visited Tahiti between 1767 and 1769, he brought home tales of people living naturally harmonious lives and of sexual freedom seemingly full of innocence. Both were attractive concepts, but that is mostly what they were—notions to seduce the minds of a continent eager to be seduced.

Had the islands contained spices or gold and been discovered in the fifteenth or sixteenth century, as nearly happened with the Spanish, Europeans would have rejected with horror the licentiousness and questionable Polynesian mores. By the eighteenth century, the mood was different and Europeans were ready to embrace new philosophical possibilities; they believed nature to be generous and beautiful and the natural man, untainted by civilization, to be a noble being.

There was much truth in the reports they heard. Indeed, the climate and vegetation of Tahiti procured an easy life for the islanders, the civilization they enjoyed seemed to lead to harmonious rapports among themselves, and there was a certain innocence in the way men and women behaved together. However, we should remember that Bougainville, the first to report on the islands and the main initiator of the Tahiti-as-Cythera myth, only stayed nine days there.

Others who gained more knowledge of these islands, such as Captain Cook and his naturalist Joseph Banks, became aware that the idyllic surface had serious cracks. The handsome and generous natives may have been noble, but they were also seen to be thieves, and could be unreliable. In other words, they were not much different from most other men. Then came the realization of even darker sides to this perfect society: intertribal warfare, human sacrifices, cannibalism, occasional infanticide, and a pitiless hierarchy. However, it was too late by then to change European perception: the myth had been adopted. Polynesian reality receded as soon as Tahiti was renamed to conform to a Greek myth and became endowed with all of Cythera's attributes.

It also turned out that the amorous licence the sailors had so much enjoyed was not as natural or universal as was first thought. As elsewhere, some women were a little more generous with their favours than others, and soon substantial trading (iron nails for sex) became commonplace. But there, too, the picture of widespread free love had been painted and was to remain firmly in European imagination. The vision of such a culture could not have pleased them more, coming as they did from a society that prided itself on being open minded, curious, literate, and well versed in Greek classics and mythical allusions.

These small Pacific islands had so far held little interest for Europeans, who had long been crisscrossing the oceans in search of material wealth. Fifteenth- and sixteenth-century explorers did not find in them what they were seeking above all: gold and spices. It is only later, with the Europeans' new fascination for scientific discoveries, that the usefulness of these same islands became apparent as convenient spots from which to observe celestial phenomena, as exotic lands full of unknown species whose discoveries could make a naturalist's reputation, or as stepping stones towards finding the Australian continent, which everyone knew ought to be nearby. Moreover, all of these islands were vitally important as the timely sources of water and fresh provisions needed to allay the sailors' sufferings and prevent their deaths from the scourge of long-distance navigation, scurvy.

Scurvy

For European fleets during those first transoceanic voyages, scurvy had been the great killer of men at sea. Caused by a deficiency in vitamin C, required for the synthesis of collagen in the human body, the disease manifested itself with a variety of symptoms, among which were spongy gums and bleeding from the mucous membranes. As the condition progressed, suppurative wounds appeared all over the body, the gums could no longer contain the teeth, and death was not far behind.

John Hale (1967:83) reports a sixteenth-century sailor's version of his dreadful plight, amply confirmed by medical descriptions. "[Scurvy] rotted all my gums which gave out a black and putrid blood. My thighs and lower legs were black and gangrenous, and I was forced to use my knife each day to cut into the flesh in order to release this black and foul blood." He also used his knife on his gums, which were livid and growing over his teeth. "When I had to cut away this dead flesh and caused much black blood to flow, I rinsed my mouth and teeth with my urine, rubbing them very hard." His case was not an isolated one. "Many of our people died of it every day, and we saw bodies thrown into the sea constantly, three or four at a time. For the most part they died with no aid given them, expiring behind some case or chest, their eyes and the soles of their feet gnawed away by the rats."

To appreciate fully the extent of the problem, we should know that scurvy killed more sailors in the eighteenth century than war, accidents, and shipwrecks combined. It is even said that captains often took on twice as many men as they needed, in prevision of the expected death toll.

Over a century later, yet matching the horrors of these earlier examples, Bernardin the Saint-Pierre, a contemporary of Cook and Bougainville, described his arrival in Isle de France (Mauritius). He came on "a ship with its flag at half mast, firing cannons every minute, with a few sailors, like ghosts, sitting on deck; our scuttles all open, out of which wafted a foul vapour; the between-decks full of the dying, the forecastle packed with the sick taking the sun and dying even as they spoke to us" (2003:89-90).

On long voyages, sailors were at the mercy of fate for provisioning with food and water in ports of call sometimes merely found by happenstance. Magellan, sailing out with a large crew of around 250, returned to Seville on the *Victoria* in 1522 with only eighteen men. Not all these deaths were due to scurvy for, as well as harsh storms at sea, the sailors met with treachery, poisoning, and violence on shore, but the records confirmed their disease and pitiful conditions. With swollen or bleeding gums, forced to eat "powder of biscuits swarming with worms" and stinking "of the urine of rats," they drank putrid water and ate "ox hides that covered the top of the mainyard" (used to prevent the yard from chafing the shrouds). "Often we ate sawdust from the boards. Rats were sold for one half ducado apiece and even then we could not get them" (Hale, 1967: 83; Gilbert, 1973: 212).

Captains did their best to provide sufficient provisions, but they seldom knew in advance the length of their voyage, and they were strictly limited by the space available on board and the difficulty of conserving food. We have the list of a seventeenth-century ship captain's standard provisions intended to feed 190 men for three months before rotting food, starvation, and scurvy took their toll: "8,000 pounds of salt beef, 2,800 pounds of salt pork, a few beef tongues, 600 pounds of haberdine [salt cod], 15,000 brown biscuits, 5,000 white biscuits, 30 bushels of oatmeal, 40 bushels of dried peas, 1 ½ bushel of mustard seed, 1 barrel of salt, 1 hogshead [large cask] of vinegar, 11 firkins [small wooden casks] of butter, 10,500 gallons of beer, 3,500 gallons of water, 2 hogsheads of cider." The captain's own store included "cheese, pepper, currants, cloves, sugar, aqua vitae, ginger, prunes, bacon, marmalade, almonds, cinnamon, wine, rice" (Hale, 1967:82).

While the provisions lasted, the sailors' diet was more or less adequate, in spite of the obvious lack of fresh vegetables and fruit on such lists. What permitted them to continue in comparative good health was finding landfalls where to restock with water, fresh fruit and vegetables, and particularly breadfruit and yams.

Unknown plants and wild fruits occasionally proved to be harmful, such as those imprudently tasted by Columbus' men in the West Indies. "Upon only touching them with their tongues, their countenances became inflamed, and such great heat and pain followed that they seemed to be mad," wrote the *Niña*'s ship surgeon. We now know that the stinging fruit was that of the manchineel tree, whose deadly poison was used by Carib Indians to tip their arrows (Hale, 1967: 58).

Even with the lucky find of comestible plants, it was often impossible to prevent scurvy or starvation. Thus, sailors were always on the lookout for whatever known or unknown islands could contribute to their diet. From the ship surgeons' point of view, any such landing could be problematic, as it often resulted in "drunkenness, diarrhea, and pox, along with wounds sustained during fights and accidents." However, "not dropping anchor risked the scourge associated with long voyages: that insidious destroyer of health, spirits, and eventually life itself—sea scurvy" (Druett, 200:141).

The improvement in sailors' diet took a long time to come. In the sixteenth century, the daily ration for crews in the British Navy was one pound of biscuit and one gallon of beer. Indeed, the Spanish Armada (1588) was repelled by Englishmen fighting on such a diet. It took Samuel Pepys, the famous diarist but also chief secretary of the admiralty (1685-88), to formulate more appropriate guidelines for naval victuals. By the time of Captain Cook, a typical day's ration consisted of one pound of salt pork, one pound of hardtack biscuit, one pound of cereal, and one gallon of beer. To supplement their basic provisions, ships were still expected to continue taking on fresh water and local foodstuff wherever they could, and these small Pacific islands, usually flush with generous vegetation and clean water, were always a welcome sight.

Nobody really was certain what caused such a disease, although poor diet was certainly thought to be part of the problem. As early as the thirteenth century, Gilbertus Acquila had recognized inadequate feeding as its cause, and in 1593, Sir Richard Hawkins had proposed that a remedy, "most fruitful for this sicknesse, is sower oranges and lemons" (Druett, 2001:143).

Sea air was also deemed to be one of the causes for the disease. "Scurvy is caused by bad air and bad food," still believed Bernardin de Saint-Pierre in the late eighteenth century. "The officers who are better fed and housed than the sailors, are the last to be attacked by this illness which even affects animals… There is no other remedy than land air and the consumption of fresh vegetables" (2003:89). Even if the officers' table was far superior to the crew's fare, preference was always given to those afflicted with scurvy, particularly when they also suffered from dysentery, as happened to Bougainville's crew in Batavia. By then, even the last little goat and the friendly puppies on board had long been eaten, and the rats were their only source of protein. Getting a good dose of land air was still thought to be a great help in curing scurvy, something they could not obtain in Batavia, where the air was foul and miasmic.

Dr. John Coulter, who believed that scurvy was caused by salt provisions, proposed the same reliance on land air (accompanied by walking on shore). On reaching the Galapagos with the *Stratford,* Dr. Coulter was successful in bringing the crew back to health. We must note that, as well as walking on the beach, the men ate a great variety of plants (Druett, 2001:150).

Bougainville, who had sailed out with oxen, sheep, pigs, and hens on board, as was customary, had also attempted to bring enough greens, including sauerkraut, salted cabbage, some lemons, chicory. He even tried to grow sorrel in barrels on deck. However, the poor quality of his provisions was no match for the length of the voyage and towards the end his sailors were reduced to eating the rats that had overrun the ships. This was, actually, the best thing they could have done, since rats have the peculiarity of synthesizing vitamin C, thus ensuring a better diet than expected for those on board.

While captains and ship surgeons struggled to prevent scurvy, there was already scientific evidence that fresh vegetable and fruit, particularly citrus fruits, helped fight the effects of Vitamin C deficiency and its devastating consequences on long voyages. One surgeon, Johann Bachstrom, urged the use of fresh vegetables as a cure (1734). Another, James Lind of the Royal Navy (1738), proved that citrus fruit was the answer to the problem. Unfortunately for the poor souls described above, Lind's recommendations were not immediately implemented.

However, individual ship surgeons on naval vessels, trading ships, and whalers would recognize over time the value of certain fruit and vegetables, and, whenever they could, brought them on board. Dr. John Lyell, on the *Ranger* in 1830, recorded the crew's improvement. "As soon as we had loaded our boats with pigs, coconuts, limes, lemons, papaw apples, breadfruit, chilis, &c., we left the shore and made the best of our way towards the ship… and spread a salubrious feast for our scorbutic crew, which would soon renovate their exhausted strength." Later, reaching Guam, he sent ashore sick sailors "to recruit their health and strength." Two of them, who had appeared close to death and could barely stir themselves a week earlier, were then "able to waddle about the deck without assistance: so powerful are limes & young coconuts in dispelling this putrid malady" (Druett, 2011:160).

Between 1780 and 1800, a number of reforms in medical staffing and victual stocking procedures took place in the English Navy, some specifically aimed at fighting the scurvy's devastation. Dr. Sir Gilbert Blane, Royal Navy Commissioner for the Sick and Wounded Board, was the first to introduce limes as prevention in 1795. Worth Estes, writing on naval medicine in the age of sail (2012) reports that the results were extraordinary: In 1795, scurvy sufferers among the sick were one in three; in 1804, one in eight; and in 1813, one in eleven.

Coming back to the crews who landed in Tahiti in the 1760s and 1770s, they, and sometimes their captains, had all suffered from scurvy at one point or another. The first to arrive in 1767, Captain Wallis and Lieutenant Carteret's crews, were particularly sick during the long passage through the Strait of Magellan. On the other hand, Captain Cook and his men did rather well, a condition he put down to a regime of "Sour Kraut" juice ordered by the ship's surgeon, Mr. Monkhouse, and to "wort" (a decoction of malt), rather than to the vegetables previously procured in Madeira and, later, in Tierra del Fuego. Actually, one may wonder why Cook relied so much on sauerkraut, rather than lime, lemon, or orange juice, much richer in vitamin C and far more palatable. It is actually believed that this preference served to delay the effectiveness of treatment for the disease in the Royal Navy (Salmond, 2003:447). The naturalist Joseph Banks, who had observed the first symptoms of scurvy in himself, had definitely opted to treat it with lemon juice, rather than with Cook's prescribed remedy (Beaglehole, 1974:170).

In spite of his preference for sauerkraut and malt, Cook was credited for having managed remarkably well to avoid scurvy on his ships. The *Endeavour* only lost forty-four men while sailing in the South Pacific. Such figures seem appalling today, but the point made by Roderick Cameron (1964:28) is that none of those deaths was due to scurvy. Thirty-three died in Batavia, either of dysentery, fever, and consumption, with the others being accidental or unavoidable, caused by drowning, freezing to death, suicide, tuberculosis, drink, and epilepsy. His record is all the more astonishing if we compare it to that of Admiral George Anson and his circumnavigation on the *Centurion* (1740-1744), a disastrous voyage on which five of his six ships were lost and 1051 out of his original crew of 1955 died, mostly from scurvy. Earlier ships fared even worse, notably an almost mythical galleon from Manila, the ghost ship *San José,* found drifting off Acapulco in 1567, everyone on board having died of scurvy (Gilbert, 1973: 268).

When all arrived in Tahiti and saw the abundance of fruit and fish so easily obtained, and when they breathed the restorative land air, they believed they have found heaven on earth. While enchanting their minds with its beauty, the island also brought their bodies back to health with a natural cornucopia from which endless food seemed to flow. The combination of these beneficial effects on their bodies and the delights of what is called by modern Tahitians "sexy loving" is undoubtedly one of the origins of the Tahitian myth.

Terra Australis Incognita

The explorers who reached the Polynesian islands in the eighteenth century had scant prior knowledge of them, even though some had been spotted as early as the fifteenth century. However, their configuration in the vastness of the Pacific Ocean was still a mystery that few were ready to investigate.

The earlier explorers—British, French, Portuguese, Spaniards, and Dutchmen—had sailed across the oceans, filled with the greed for spices and gold. They were driven by kings and merchants, and sought access to the riches of the East. The Pacific islands seemed to lack any economic appeal, and their discovery, mainly by Spanish navigators, did not seem very important. The sailors' interest was definitely elsewhere. "We Spaniards," said the explorer Hernando Cortes, "suffer from a disease that only gold can cure" (Hale, 1967:105). Columbus is reputed to have said, upon seeing Cuba on October 27, 1492, "Where there is such marvelous scenery, there must be much from which profit can be made."[1] Their discovery of the New World was the accidental consequence of this quest. The same Europeans, while desperately seeking a shorter and easier route to Asia, had also developed an interest in *Terra Australis*. By the end of the Age of Discovery, most continents had been charted and partly colonized, and the limits of the world were assumed to be fairly well known. Yet, there was a strong sense that there was another continent somewhere in the southern part of the Pacific Ocean, which scientists, navigators, and mapmakers alike had dubbed, for lack of precise knowledge, *Terra Australis Incognita*, or even *Terra Australis Nondum Cognita*, so sure were they of its existence. Europe was eager to discover it.

Intuiting the presence of this continent made eminent sense for two reasons. The first, perhaps not entirely scientific by our standards, was the abhorrence of a vacuum as Europeans faced the immensity of the Pacific Ocean. Since the time of Ptolemy, they had suspected that there ought to be some weighty landmass to counterbalance the enormous bulk of the northern hemisphere. A more harmonious vision of the world required some equal terrestrial presence in the southern hemisphere in addition to Africa, a notion that was not entirely satisfied with the later discovery of South America. A third continent would be required to obtain the desirable balance.

The second reason was that they already knew there were islands strewn into that part of the ocean. They had no notion of the extent of the various archipelagos and the enormity of the territory that would later be discovered and grouped in 1831 into three distinct areas by the French explorer

Dumont d'Urville. These were named Melanesia (the "black" islands), Micronesia (the "small" islands), and Polynesia (the "many" islands), in a nomenclature presented in a paper for the *Bulletin de la Société de Géographie* in Paris, that still remains the one used today. People definitely knew there were islands there, and many believed these islands to be part of a larger landmass. Quiros, the influential Spanish explorer, believed them to be "the children" of a "mother continent."

Several navigators had already stumbled upon some of these islands, usually by accident, while on their way elsewhere.[2] Among them was Magellan, who named the vast expanse of sea "Mar Pacifico" and came across Pukapuka in the Tuamotu and Gambier archipelagos before being killed in Cebu in 1521. Another, the Spaniard Luis Lopez de Villalobos, sent to follow up on Magellan's discoveries in the Pacific, came upon other islands in 1542 while attempting to take possession of the Philippines. Yet another, Alvaro de Mendaña y Castro, sailed on a 1567 expedition between the Tuamotus and the islands he would later name the Marquesas [Las Islas Marquesas de don Garcia Huntado de Mendoza de Cañete, to give them their full name], in honour of his patron and uncle, the viceroy of Peru. They would occasionally stop for food and water at some of those islands, which usually held no other interest for them. After Mendaña's return and his negative report, the official in charge of informing King Ferdinand of Spain on the potential merits of new territories wrote, "In my opinion… the islands they discovered were of little importance, although they say they heard of better lands; for in the course of these discoveries they found no specimen of spices, nor of gold and silver, nor of merchandise, nor of any other source of profit, and all the people were naked savages." The same writer called them the Solomon Islands, even though they did not appear to hold any of King Solomon's famed mines (Whitefield, 2000:130).

In spite of the official lack of interest expressed by the Spanish authorities, Mendaña was keen to go back to the Solomons, explore them further, and colonize them. He was not able to return until 1595, this time with a rather unusual complement of 378 people on the *San Jeronimo*, the *San Ysabel*, and two smaller ships. Intending to colonize the island, he took along sailors, soldiers, private citizens, priests, and others who might constitute potential settlers. He also had on board his quarrelsome wife, Doña Ysabel Barreto, and her three brothers, one of whom captained the flagship. Fortunately for the expedition, the pilot was the reliable Pedro Fernandez de Quiros. It was an ill-fated voyage and a disastrous expedition that included serious clashes with natives, the disappearance in the fog of the *San Ysabel*, and malaria on board. Mendaña himself died of the sickness, together with many men, women, and children. What was left of the expedition finally returned to Manila under Quiros's guidance.

The same Quiros was given the command of four ships to return to the Pacific in 1605. They left from Callao with three hundred soldiers and sailors, ten monks to look after the sick, and no women on board. They reached the Tuamotu Islands in 1606, then another island in the northern Cook group. Finally, they arrived at an island he assumed to be *Terra Australis* and named it Australia del Esperitu Santo (later renamed New Hebrides). The expedition had not fared very well either, with several clashes with the natives, and the ships returned to Acapulco at the end of 1606. A captain of one of those ships, Luis Vas de Torres, unknowingly came even closer to the coast of Australia, and was probably the first European to see the new continent in 1606.

This was to be the end of the great Spanish discoveries and of the Spaniards' interest in that part of the Pacific. Portugal had been maintaining her grip on the Moluccas, and Spain decided to focus her efforts on establishing more securely her hold over the Philippines. By the eighteenth century, Peru had taken over from Spain the continuing exploration of the Pacific.

The Dutch, among the fearless navigators of the time, did not lag behind in exploring the Pacific, with Abel Tasman coming upon the islands of Rotterdam and Amsterdam (today's Nomuka and Tongatapu) in 1595, then sailing before the westerlies to discover in 1642 Van Diemen's Land (Tasmania) and New Zealand, while sighting Fiji along the way. Nearly a century later, Jacob Roggeven, searching for *Terra Australis* on behalf of the Dutch West India Company and finally heading back for Batavia, encountered the Tuamotu archipelago, the Society islands, and Samoa.

Such was the situation during the Age of Discovery, a period when Europeans repeatedly sailed across the Pacific. They sighted and often landed on some of the myriad islands scattered throughout the ocean, sometimes finding them by accident, sometimes seeking them specifically, as did for instance Mendaña and Quiros.

The eighteenth century was different. Europeans sought these islands in earnest, even if often as a mere stepping stone to the eventual discovery of Australia, their ultimate goal. With the exception of the Spaniard Boenechea from Peru and the Frenchman Bougainville, they were English and they all stopped in Tahiti.

By the time of James Cook's first voyage (1768), it was generally accepted that, somewhere south of latitude 50°, east of Africa, and west of Cape Horn, there was a continent whose presence everybody suspected. Some came very close to sighting it. Many had been inspired and guided by Quiros, who believed the hidden continent to be just to the south of Mendaña's route. Bougainville, leaving Tahiti in 1768 and still curious about the southern continent, sailed due west, and ran into heavy breakers and considerable rocks and reefs that forced him to change his course to the northeast. Unbeknownst to him, he was close to the Great Barrier Reef. However, not recognizing what Quiros had described as the approaches to *Terra Australis* and weighing the dire conditions on board (near starvation and scurvy), he was forced to abandon further exploration and return in haste to safety.

In spite of the general belief in the continent's existence and some frustrated navigators' best efforts to discover it, none of this first group reached Australia, and it was not until 1775 that James Cook solved the mystery of the new and unknown landmass. By then Tahiti had also been explored and charted.

The New Explorers

Jean-Jacques Rousseau deplored the general ignorance in Europe of previously discovered lands and attributed it to the nature and character of the navigators involved. At the time, voyages of exploration had been mostly undertaken by sailors and merchants, rather than scientists and artists, who would have documented these new lands and populations. Indeed, given the mercantile nature of these earlier voyages and the intense competition among countries, secrecy, misdirection, and misinformation would likely have been taken for granted. More to the point, these navigators had no particular interest in reporting on geographical considerations and on native customs, as those would have been outside their remit.

However, the eighteenth century, with its great intellectual currents awash across Europe, was curious about the world at large. Inevitably, the earlier lack of substantial information and poor reporting would be amply remedied by those who landed in Tahiti between 1768 and 1793. No longer exclusively driven across the oceans by their obsessive quests for wealth, many of these new navigators were bent on scientific missions. Others on board—scientists, naturalists, surgeons, artists, and even midshipmen—were educated men who also reported in writing on their journeys to the enchanting island. How this change came about is worth describing briefly since it arose from different economic and scientific motives in discovering new lands and corresponded to a totally new frame of mind among those who sponsored, organized, and led expeditions.

The following is the list of those navigators who successfully reached Tahiti, each with a particular purpose and mission drawing him to the South Seas.

Navigators/Ships Nationality	Purpose/Event	Crew and Others
Samuel Wallis (1767) *Dolphin & Swallow* English	Search for Australia	Phillip Carteret on Swallow
Bougainville (1768) *Etoile & Boudeuse* French	Circumnavigation. Mission to Falklands.	• Naturalist Philibert Commerson & Jeanne Barré; La Pérouse; astronomer Véron
James Cook (1769) *Endeavour* English	1st voyage (1768-71) Transit of Venus. Search for Australia.	• Naturalists Joseph Banks, Solander & Spöring. • Astronomer Green • Artists Parkinson & Buchan

James Cook	2nd voyage (1772-75)	• Artist William Hodges
Resolution & Adventure	(1st chronometer) Tahiti, Easter Islands, Marquesas, Tonga, New Hebrides.	• Botanist Anders Sparrman • Captain Tobias Furneaux on *Adventure* • Bligh, second master
D. de Boenechea (1772) *Aguila* Spanish	Exploration (from Peru) on behalf of Spanish crown	
Boenechea (1774-75) *Aguila & Jupiter*	Establish mission. Died in Tahiti (1775)	Catholic missionaries
James Cook (1777) *Resolution & Discovery*	3rd voyage (1778-79) Hawaiian islands 1778. Died on Kealakelua in 1779	•William Bligh on *Resolution* •George Vancouver •Naturalist Nelson •Astronomer Bayley
William Bligh (1777) *Resolution* English	On Cook's last voyage. Hawaii/ Tahiti with Cook	see above
W. Bligh (1788-89) *Bounty*	1st bread-fruit expedition. Mutiny.	• Fletcher Christian •Naturalist Nelson
W. Bligh (1791-93) *Providence*	2nd (successful) breadfruit expedition.	• Botanists James Wiles & Charles. Smith

Eighteenth-Century European Navigators' Visits to Tahiti

In addition, two other ships arrived in Tahiti soon after the *Bounty*'s mutiny: The *Lady Perhyn,* with Captain Sever (July 1788), a prison ship back from Botany Bay with her small crew, sick with scurvy, getting fresh food and restoring their health, and the *Pandora* with Captain Edwards (end of 1788), sent to Matavai Bay to hunt down the *Bounty*'s mutineers.[3] These two visits were significant in that they augured the reawakening of the Europeans' interest in Tahiti and their return to the island after the interruption of the American War of Independence (Salmond, 2009: 454). The focus of interest had shifted to scientific discoveries and most ships now had naturalists on board. Their reputations have often survived through their discoveries and the names given to the various species they were the first to identify, celebrating their own fame or that of their patrons.

Even before the most famous of them, Charles Darwin, spent five years on the *Beagle* investigating the distribution of wildlife and fossils, thus leading to his theory of evolution, others had left their mark and made a lasting name for themselves. Two were directly associated with Tahiti: Joseph Banks, who accompanied Cook on his first voyage, and Philibert Commerson, who sailed with Bougainville. Their own impressions, confirming those of their captains, added sufficient scientific weight to the general depiction made of Tahiti that it would eventually have been taken seriously, if perhaps not when first reported in Europe.

Thanks to this congruence of reports, sojourns on Tahiti are well documented. Within a quarter of a century Wallis (1767), Bougainville (nine months later in 1768), Cook (a year later in 1769, then again in 1777, with a young Bligh to accompany him), and Bligh again (1789 and 1793) all spent time on Tahiti. Others went there as well during the same period: the Spaniard Boenechea, from Peru, and the two British prison ships already mentioned. The gist of the reports from Tahiti came from those on board the *Boudeuse* and *Etoile*, the *Dolphin* and *Swallow*, the *Endeavour*, the *Resolution*, the *Discovery*, and the *Bounty*.

Very likely, they had not expected to find such a bountiful island when they reached Tahiti, nor had they expected to be so utterly seduced by the land, the people, and the culture they were discovering. Some had landed in Tahiti almost accidentally as part of a larger voyage of discovery, as did Bougainville; others had a specific scientific motive for being there, including astronomy for Cook and botany for Bligh. In neither case were they in Tahiti to study the population, as would later variously be the case with missionaries, anthropologists, travel writers, or novelists. Yet, they left a most extensive record of the vitality of the culture they encountered there. They studied the Tahitian geography, habitat, flora, fauna, language, tattoos, manners of dress, religious rituals, and the reports they wrote constitute the basis of many studies for future centuries.

Tahiti happened to be the best location from where to observe the transit of the planet Venus, the mission with which Cook had been entrusted. Edmund Haley, the Astronomer-Royal, had recommended that the next observation of this rare event, due to occur on June 3, 1769, be used to determine the solar parallax and calculate the scale of the solar system and the distance between the Earth and the Sun. An earlier attempt in 1761 having failed, it was all the more urgent to try again eight years later, since the transit occurred infrequently and with little regularity. Cook was chosen to lead the mission on the *Endeavour* under the aegis of the Royal Society, with Tahiti as the evident destination. Wallis, upon his return, had recommended the island as the optimal site for the observation.

Unlike many South Seas islands, Tahiti's position was by then accurately marked on the charts, even if there was still certain amount of secrecy in the matter of map-making. Both Cook and Bougainville kept the details of their charts and directions to themselves, so when the Spanish expeditions sent by the viceroy of Peru[4] between 1772 and 1775 to explore and evangelize Tahiti, they had no sailing instructions they could trust (Salmond, 2008:15). Domingo de Boenechea, their captain, had to rely on native navigators as they sailed though the Society Islands, the Marquesas, and others—as had other Europeans before them. Such guesswork would not be the case with Cook's visit to Tahiti. The island had been well charted by the English and a few of Wallis's former seamen on the Tahiti voyage were also on Cook's ship. In keeping with the century's new interests and the scientific nature of the mission, *Endeavour* also carried on board the naturalist Joseph Banks and his two Swedish assistants, Daniel Solander and Herman Spöring. One of Banks's friends was of the opinion that "No people ever went to sea better fitted out for the purpose of Natural History. They have got a fine library of Natural History; they have all sorts of machines for catching and preserving insects; all kinds of nets, trawls, drags and hooks for coral fishing" (Salmond, 2008:80).

There were also two artists on that first voyage: Alexander Buchan, who was to paint figures and landscapes, and Sydney Parkinson, the local fauna and flora. Buchan's premature death soon upon arriving in Tahiti left only William Hodges, who sailed with Cook on his second voyage, to provide a pictorial record of Tahitian daily life, artifacts, and landscapes during the early days of contact with the Europeans. They became particularly important in helping reintroduce later in Tahiti the evidence and details of an earlier sophisticated culture.

Bligh's mission was equally specific: to collect and transport the Tahitian breadfruit to England and then to Jamaica, where it would serve to feed cheaply the slaves on sugar plantations. The Europeans believed, rightly as it turned out, that the providential breadfruit tree could be transplanted to their own colonies with a similar climate to Tahiti's. The French had been on a similar quest and were successful in transplanting it in 1793 from Java to the Ile de France (Mauritius). The value of the breadfruit to colonial powers was inestimable. Joseph Banks wrote of the breadfruit that if a man could plant ten trees, taking only an hour of his time, he would have fulfilled his duty to future generations.

The delicacy of Bligh's mission resided in having to wait for the trees to be seeded, and to nurture the saplings to a point where they could be safely stored in a makeshift nursery on board. His first attempt in 1788 failed dismally, for reasons that are only too well known. Released at sea with his botanist David Nelson and a skeleton crew after a mutiny on the *Bounty* led by Fletcher Christian, Bligh achieved the navigational feat of returning to England by covering the 3618 miles to Batavia in 47 days. He lost only one sailor to the natives of the Loyalty Island and five more to disease in Batavia (including the unfortunate Nelson). Whatever his failings, he was an incomparable sailor.

Charged again with the same mission by Joseph Banks, then patron of Kew Gardens and president of the Royal Society, Bligh returned to Tahiti on the *Providence* in 1791 with two gardeners, James Wiles and Christopher Smith. This time, he successfully brought back sixty-six breadfruit specimens.

While Cook, Boenechea, and Bligh's specific destinations had been Tahiti, the others found themselves there often by chance. Wallis was actually looking for Australia, and the French Bougainville was on his way to negotiate with the Spaniards the establishment of a French territory on some strategic islands off the coast of Argentina, variously known as the Malouines by the French, the Malvinas by the Spaniards, and the Falklands by the British. Bougainville would then proceed on to his initial mission, a circumnavigation of the world, and narrowly miss the Great Barrier Reef in so doing.

Among those who participated in more than one expedition, Cook and Bligh were experts in providing careful descriptions of what they had seen on Tahiti. Professor Salmond, who has written extensively on the subject, particularly deplores the loss of Bligh's journals and maps during the *Bounty* mutiny, recognizing his remarkable qualities as a pioneer ethnographer of the early contact period. We should remember that, even before the foundation of the Royal Society in 1660, an established tradition of the British navy had been the empirical observation of nature, as could be found illustrated in detailed ships' logs and seamen's journals. It is this long-established tradition of empirical observation—not just in the British Navy, but the French one as well—combined with the scientific training of the naturalists on board, that produced the invaluable reports of Cook, Banks, Bougainville, Commerson, and others.

Although much of Bligh's original material is lost, he remains one of the most important contributors to the knowledge of events that took place during that period. He figures as an essential witness to two very disturbing events: the mutiny on his ship during the first breadfruit expedition—for which he blamed Tahiti, "the Paradise of the World," and "the allure of the island women"—and Cook's slaying in Hawaii, both of which required clarification. The latter event, in particular, left Bligh bitter with what he felt was its misrepresentation of the facts by Cooks' officers (Salmond, 2011: 20, 41-44).

Both the mutiny and Cook's death within a decade would have been enough to draw attention to Polynesia and the natives of those Pacific islands, so recently discovered. Combined with the unanimous description of earthly delights to be found there, Europe was ready to be interested and eventually seduced.

Few have spent as much time in Tahiti as Bligh. Both as a sailor on three visits, surveying and taking notes for his patron Joseph Banks and, later, as a semi-fictional character featured in books and films, he exemplifies the European presence in Polynesia and is, indirectly, a major contributor to the creation of the Tahitian myth.

When we think of the harsh conditions at sea and the prevailing threat of scurvy on long voyages, we can well imagine that any landfall with fresh water and abundant fruit was seen as the answer to the captain and the crew's prayers. Moreover, when the natives were friendly enough to procure those willingly and generously, the sailors may well have thought themselves blessed. When the island they visited was Otaheite, they thought they had arrived in paradise.

Polynesian Seafarers

Well before the eighteenth-century expeditions brought the Polynesians to the attention of Europe, there had been another type of explorer in the same ocean. The Polynesians themselves were expert boat builders and superb navigators in search of new islands to inhabit. Their explorations over the Pacific and their navigational feats astonished the English, the French, and the Spaniards, famous for their own daring crossings.

There are many speculations on the origins of Polynesian migrations and expeditions, but I draw from Anne Salmond (2003: 35) to sketch the following outline. The Polynesians had very likely started their original journey from Taiwan across the Pacific five millennia ago, in a slow island-to-island migration, before finally reaching Polynesia about a thousand years BCE, when the first Lapita settlement was recorded in Tonga. Then, fairly late in the Christian era, between the 1100s and the 1300s, they sailed again from their Pacific homeland, variously named Hawaiki by the Maoris, Havai'i by the Society Island natives, Savai'i by the Samoans, or simply Hawai'i. This origin name does not actually refer to what is now known as Hawaii, and it is difficult to trace with exactitude their progress throughout the Pacific Ocean because of their habit of giving old names to new lands. They are thought to have first reached Samoa, Tonga, and other life-supporting islands to the north. Later (the dates are equally vague) they are believed to have pushed in three different directions: south to the Austral Islands and New Zealand, north to Hawaii, and east to Mangareva and Easter Island,[5] forming a vast triangle. The location of the part we now call Polynesia roughly extends between the Tropic of Cancer to the North and the Tropic of Capricorn to the South, and between 150° (West) and 120° (East), a territory larger than Melanesia and Micronesia together.

French Polynesia is only a relatively small part of this vast expanse. It is constituted of six archipelagoes: Tahiti, more or less at its centre and with the largest and capital city, Papeete; Moorea; the Leeward Islands, the Austral Islands, the Tuamotu and Gambier Islands, and the Marquesas Islands.

With such an expanse of water and the multitude of small islands scattered in it, it is not surprising that blue water sailing is thought to have started in the Pacific Ocean, when groups of people first set forth on bamboo rafts from Southeast Asia towards Australia and New Guinea fifty thousand years ago. Their descendants became the first Pacific Islanders, sailing to other islands, bringing their own traditions, their language, their myths, their implements, their animals, their cultivation techniques, and their building styles. When Cook discovered Hawaii, he found to his astonishment that the natives, while slightly darker of skin than the Tahitians, spoke a similar language, and he remarked, "How shall we account for this nation's spreading itself so far in this vast ocean?" (Gilbert, 1973:300).

Cook had already seen the Tahitian canoes' performance and heard one of his crew estimate that a Tongan canoe could sail "three miles to our two," so why should he have been so surprised at their wide-ranging sailings? In his review of the settlement of Polynesia, Kawaharada describes the natives' outrigger or double-hulled canoes. Built of tree trunks and carved with adzes, or made from boards linked together with ropes of twisted and braided coconut fiber, they had often proven their sea worthiness to the Europeans. Sails made of woven coconut or pandanus leaves were used when the wind blew, and powerful oarsmen paddled the canoes when the sea was calm. For long voyages, rather than using single canoes with outriggers attached for greater stability, the Tahitians would lash two canoes together with crossbeams and a deck spanning the two hulls.

The natives of Raiatea in particular were reputed canoe builders. They were known to use breadfruit trees for timber, which are impervious to salt-worms. Some of these canoes, built with stone adzes and chisels made of human bone, could carry up to three hundred people. Banks was particularly fascinated with the boat builders' dexterity, seeing them hollowing "with stone axes as fast at least as our carpenters could do, and dub, though slowly, with prodigious nicety" (Cameron, 1964:122).

Ignacio Andia y Varela, a Spanish contemporary of Cook who had charted the coast of Chile, had the opportunity of closely observing these canoes. He described his visit to the Society Islands in 1774 and wrote of his admiration for the canoes he saw there.

> It would give the most skilful [European] builder a shock to see craft having no more breadth of beam than three

[arm] spans carrying a spread of sail so large as to benefit one of ours with a beam of eight or ten, and which, though without means of lowering or furling the sail, make sport of the winds and waves during a gale, their safety depending wholly on two light poles … (about eight feet) long, which, being placed … the one forward and the other aft, fitted to another spar of soft wood place fore and aft wise in the manner of an outriggers. These canoes are as fine forward as the edge of a knife, so that they travel faster than the swiftest of our vessels; and they are marvelous, not only in this respect, but for their smartness in shifting from one tack to the other. (Corney, 1913:282)

The Polynesians' success in reaching new islands over enormous distances and facing immense dangers may have been due to chance. However, the islanders seemed to have been well prepared and set out with men, women, children, their pigs and dogs, edible rats, some seeds and shoots, sugar cane, yams, and mulberry seeds for making bark cloth, ready to settle on whatever new lands they would come upon. They used paddles on calm days, and sails when the wind blew. They had faith in their knowledge of the sea and sky, and in the support of their gods.

The Europeans were also impressed with the Tahitians' extraordinary ability to orient themselves around their ocean. This admiration was merited when we compare the two types of navigational aids. The eighteenth-century European seamen who discovered Polynesia and were able to find their way there again with great precision could rely on the accuracy of their celestial navigational instruments.

Earlier on, graded sights had to be aligned with the sun or a star and astrolabes were used in this rough calculation. While there were tables for measuring the angle between the horizon and the North Star at night or the sun at noon to determine how far north or south of the equator the ship sailed, much depended on the accurate calculation of the time of the observation, which was not always possible to obtain at sea, particularly given some captains' rather basic mathematical knowledge.

The compass had long been the most reliable instrument available to navigators, enabling them to sail beyond the sight of land. When on long voyages, the custom was to sail along the same latitude, once it had been established through celestial observation, in spite of the necessary tackling when the wind required it. For those sailing from Europe to the West Indies, the technique was summed up as "south till the butter melts, and then due west."

Some innovations permitted better calculations. The quadrant, credited to Pierre Vernier in the 1630s, helped navigators in determining the altitude of the sun above the horizon. One of the serious problems was that the ships' clocks did not keep the right time at sea. However, by the middle of the eighteenth century, two inventions had made navigation far more accurate. John Harrison, a clockmaker, developed in the 1730s, after years of trials, the first practical marine chronometer, thus enabling sailors to calculate their longitudes. About twenty years later, the mathematician John Hadley invented the octant, a precursor of the sextant, which permitted adequate measurement of the altitude of the sun or any other celestial object above the horizon at sea.

It is from this viewpoint of modern eighteenth century navigation that the Europeans assessed the merits of the Tahitians' progress at sea. The latter are believed to have marked out the ocean by sea-paths between recognizable features, star configuration at night, sun position during the day, landmarks among the cluster of known islands, and any other clue they could gather from the observation of nature.[6] Among those particularly adept at navigation, through his thorough knowledge of sea currents, astronomy, and meteorology, was Tupaia, originally from Raiatea, who had come to Tahiti to flee the wars on his island. There, he quickly rose through the ranks of the elite *arioi* and became well known for his many talents. He met Wallis, Banks, and Cook, for whom he drew an accurate map of Polynesia, naming seventy-four islands and aware of nearly twice as many. He so impressed Cook that the latter asked him to guide the *Endeavour* through the archipelagoes and to New Zealand.

Through him, Cook had noted that Tahitian navigators used the rising and setting points of the stars to orient themselves at sea, but Andia y Varela was given a more specific description of the manner they calculated this path. "They have no mariner's compass, but divide the horizon into sixteen parts, taking for the cardinal points those at which the sun rises and sets… The helmsman … knows the direction in which his destination bears. He observes also whether he has the wind aft, or on one or the other beam, or on the quarter, or is close-hauled. He notes, further, whether there is a following sea, a head sea, a beam sea, or if the sea is on the bow or the quarter. He proceeds out of port with a knowledge of these [conditions], heads his vessel according to his calculation, and aided by the signs the sea and wind afford him, does his best to keep steadily on his course."

Cloudy weather made the navigation more difficult, as did changes in the condition of the ocean. But their knowledge of nature and its phenomena was so profound that they succeeded in keeping track of their position at sea during the long voyages to distant islands. They memorized the distance and direction traveled and could anticipate the presence of islands before actually seeing them. "Voyagers followed the flight of land-dwelling birds that fished at sea as these birds flew from the direction of islands in the morning or returned in the evening. The navigators also watched for changes in swell patterns, clouds piled up over land, reflections on clouds from lagoons, and drifting land vegetation."

Andia y Varela had been equally impressed with the Tahitians' unparalleled ability to read and anticipate the weather. "Every evening or night, they told or prognosticated the weather [for] the following day as to winds, calms, rainfall, sunshine, sea and other points, about which they never turned out to be wrong: a foreknowledge worthy to be envied for, in spite of all that our navigators and cosmographers have observed and written about the subject, they have not mastered this accomplishment" (Corney, 1913-1919, Vol. II:284-87).

Their ocean held a mystical character for early Tahitians, comparable to that of a *marae*, the sacred place of their spiritual and social life. They felt guided by their gods and their ancestors in their maritime quest. Polynesian poets have sung of their epic, heroic expeditions, voyaging under the sky and seeking new lands in their large outrigger canoes built with tools of stone, bone, and coral.

> My heart goes back to the beginning of time / Here I will ask /
> Where did I come from? / My ancestors came from 'Avaiki /
> They sailed down to the south / Tangi'ia nui, with 8 times 200 people /
> Bestowed the name Rarotonga / At Avarau and Vai-kokopu /
> My canoe landed / Vaerota was also a port of call /
> And a *marae* for my ancestors / This will remain an example /

For the generations in the future / This will remain an example / Forever and ever (Te Ariki-tara-are).

Another poem sings of another arduous journey:
>I am Uke Ariki / Who has travelled on the Blue Great Ocean./
>The clear skies of Tangaroa, / The clear skies of eastern Avaiki, /
>The moving of the sun, / The haloing of the sun, /
>The nights of the new moon, / The nights of the full moon. /
>As a high chief, I have arrived at … Akatotakamanava.
>The travelling of the canoes, with three hundred and eighty…
>I have arrived in the shade of the peaceful land of Avaiki-tau-aro
>Came ashore… / At the harbour of 'Uke Ariki /
>And I settled in the district of 'Itaki and built the *marae* of
>Rangimanura.
>(Akatokamanava)

They risked starvation, capsizing or swamping their canoes in heavy seas, having their sails ripped apart, their masts or booms broken by winds, smashing their hulls against underwater rocks. They bore the bad weather and cold temperatures at night with only capes made of leaves or bark-cloth to protect themselves. The chants celebrate mostly the successful arrivals, promptly followed with the building of a *marae*, the Tahitians' most symbolic action to take true possession of the land.

Teuira Henry (1928:150) explains what they meant to her people. "Marae were the sanctity and the glory of the land… the pride of the people… the ornaments of the land… [and] the palaces presented to the gods." They were usually adorned with *unu*, the carved panels and posts of religious and symbolic significance that were placed in the middle of the *marae* or directly in the front of the *ahu* platform at the head of the *marae*.

Restored Taputapuatea *Marae* and *Unu* on Raiatea

The chants usually celebrate successful landing on a new island, but catastrophic voyages are known to have occurred as well. One such voyage from Hiva in the Marquesas to Rarotonga is recorded by Kawaharada as follows: "The voyage was so long; food and water ran out. One hundred of the paddlers died; forty men remained."

Although disasters did occur, the explorers may have been more cautious than is usually supposed. They may have adopted a less risky strategy of exploration by waiting for a reversal in wind direction and sailing upwind (eastward in the Pacific) as far as their supplies lasted, then returning home when the wind shifted. This tradition, called '*imi fenua* or "searching for land," is reported in the Marquesas and other Polynesian islands.

Whether exploration was practiced in this manner or in a different, riskier one, the Polynesians, mostly Tahitians and Marquesans, did spread their influence over a vast expanse of the Pacific Ocean, with the Hawaiian Islands becoming their most important settlements.

Hawaiian scholars attribute the discovery of Hawaii to a fisherman called Hawai'iloa on a fishing trip from his homeland of Ka'Aina kai melemele a Kane ("Land of the yellow sea of Kane"). According to this tradition, the Big Island was named after him, and the other Hawaiian Islands were named after his children. Elders interviewed in the early part of the twentieth century believed this origin myth to be their reality. This historical legend is revisited in today's naming of some of the Polynesian Voyaging Society's canoes.

Another myth, transmitted by Teuria Henry, attributes the discovery of Hawaii to another hero, this one named Tafa'i, who fished up the islands from their ocean bed. He failed in his attempt to draw the Hawaiian Islands closer to Tahiti and her own islands, but the myth illustrates the close Polynesian connection between fishing and discovery, and suggests that Tahitian fishermen may often have been among the original travellers and discoverers of new islands.

Outrigger canoes were essential to the life of the islanders and participated in all affairs of importance, including transportation, exploration, fishing, war, and entertainment. They were the superbly efficient means of exploration we have seen, but also had enormous fighting power. Their upward curving prows were impressively decorated with mythological characters, gods, or protective spirits. In Cook's time, there were in Tahiti

160 such canoes, as well as an even larger number of smaller vessels, constituting an impressive outrigger fleet that was successfully used to subdue rivals at sea. However, these canoes were no match for the much better equipped ships of the Europeans, even if these Europeans recognized the outriggers' fighting ability during the skirmishes that opposed them to their own vessels.

War Canoes in Tahiti (William Hodges)

Beside exploration and warfare, the canoes had other important social functions, such as carrying the *arioi* religious dancers and artists to their various ceremonial performances across the islands. Those journeys were no modest affairs, both in numbers and in extravagance, corresponding to the importance and popularity of the events. J. A. Moerenhout, who resided in Tahiti in the 1830s and had sailed by small ship through Polynesia, reported hearing of "flotillas of a hundred and fifty canoes coming at the same time from Raiatea, Huahine, etc., each having on board seldom less than thirty or forty, and sometimes a hundred persons" (1837:132). While those numbers seem high, Forster, travelling with Cook, reported seeing an *arioi* flotilla of seventy canoes at Huahine and believed the number of people on them being as high as seven hundred. Such flotillas would be adorned with flowers, and the *arioi* themselves would be resplendent in yellow-leafed girdles and red coats, while naked dancers on platforms followed the rhythm of the sharkskin drums. They were led in great ceremony by their grand master wearing a huge adornment of red feathers on his head. Such arrivals would gather the whole population on the shores, where a great show of respect and admiration would be made while greeting the *arioi* performers. Their arrival would lead to exciting events, including sexually explicit dancing, lovemaking, kava drinking, and drama performances (Salmond, 2003: 256).

Such were the sea-faring traditions and the well-developed coastal and maritime culture the Europeans discovered when they reached Tahiti.

[II]
OTAHEITE, EARTHLY PARADISE.

WHEN LOUIS-ANTOINE, COMTE DE BOUGAINVILLE, first laid eyes on the idyllic island of Otaheite in 1768, he had to rename it *La Nouvelle Cythère*. It is true that exotic islands have a universal appeal. Cecil Arthur Lewis wrote in *The Trumpet is Mine* that all his life he had longed to go to the South Seas, and describes what we all have in mind when the same thought possesses us. "The very words had magic in them, conjuring up primordial memories of palms and coral reefs and sun-drenched sands; of fruits and flowering trees; of women warm in welcome; of peace and isolation; a shore where the struggle for existence is unknown and where, for once, the rose proliferates without a thorn."

Would any South Seas island satisfy these cravings? To some extent, many of them would fit Lewis's description, particularly in the 1930s, when the book was written, and when the rose had probably fewer thorns than are now to be found. But Tahiti is in a special category—and not only because it was extensively and lavishly painted by Gauguin. Much has been written about its emotional and sensual appeal, and these writings initially contributed to the elaboration of the fertile Polynesian myth of a benevolent and innocent nature, the physical beauty of the men and women who inhabited it, as well as their sexually accommodating nature. The tone was set when the Europeans first landed on Polynesian shores, and it would be hard to deny that these islands occupy a privileged place in Western imagination.

There was something distinctly special about them. We should remember that the *Nuremberg Chronicle* of 1493, geographically obsolete as soon as it was published since it did not include the map of the Americas, depicted rather fabulous and sometimes horrifying beings inhabiting unknown lands. Inspired by classical mythology and sailors' yarns, it created images of what "monsters" might lie within. The *Chronicle* is part of a long tradition of demonizing the unknown, and legends survived because Europeans knew so little then about the continents that might hold such wonders.

It is unlikely that sailors entirely believed these myths by the end of the eighteenth century—even if Bougainville's men still believed that the Patagonians were seven-foot tall giants with huge feet. However, there must have been, in spite of the big guns their ships carried, a great deal of nervous tension and perhaps superstition when approaching islands no Europeans had ever visited before. There had already been many violent confrontations between sailors and natives of newfound lands, even when these natives looked like normal human beings. The beauty and welcoming disposition of the Tahitians must have been a most welcome surprise, though skirmishes did occur between islanders and newcomers. Thefts were common, but on the whole the impression left was that of an idyllic land.

This is particularly true when we compare Tahitians with the inhabitants of other lands just discovered. As was the fashion of the times, the Tahitians were often referred to as "Indians" or "savages" and described as "primitives," terms which, according to Fernand Braudel, are inappropriate to describe the Tahitian culture at the time of contact. Not only had they designed the efficient outrigger canoes, but in spite of the abundance of wild plants and fruit surrounding them, they cultivated yams, sweet potatoes, gourds, and sugar cane. They also reared pigs and poultry to suffice to their needs. Braudel then quotes from Wallis to show what true primitive people actually were like. First met by Cook in Le Maire Strait, at the outermost tip of America, were a handful of "wretched savages, deprived of everything, and with whom the Europeans were unable to make real contact." Two years earlier, Wallis had met and reported on the same savages. "[One of our sailors] who was catching fish, gave to one of these Americans a live fish which he had just caught, a little larger than a herring. The American seized it as avidly as a dog does a bone that is thrown it: he killed it first by biting it near the gills, then began to eat it starting with the head and proceeding to the tail without discarding the bones, fins, scales or guts" (Braudel, 1982:180).

By the nineteenth century, Tahiti's reputation had been well established. New diseases, missionaries, traders, whalers, slavers, and planters had already left their marks, but the fascination still endured. The second wave of those who wrote about the island consisted mostly of adventurers, novelists, administrators, or artists. Their vision had occasionally shifted to include some reservation and a more realistic appreciation of the native culture, yet retained much of the previous era's sense of wonder. Among them figure Pierre Loti, Hermann Melville, Robert-Louis Stevenson, and Paul Gauguin.

A later group of individuals writing about Tahitian history, culture, and society—some our contemporaries—provide a wider scope. They include anthropologists, sociologists, linguists, novelists, artists, cinematographers, historians, politicians, musicians, poets, civil servants, and social activists. Among so many others we find Rupert Brooke, Zane Grey, Jack London, Pierre Loti, William Somerset Maugham, Alan Moorehead, Bengt Danielsson, Robert Turnbull, Jacques Brel, Anne Salmond, Bruno Saura, Miriam Kahn, and some Tahitian novelists, such as Chantal Spitz, Célestine Vaite, and Michou Chaze. Many, particularly among the latest writers, have a different perception of the island, its past, and its destiny.

By casting our nets a little wider, we naturally find all those who have written extensively about other Polynesian islands with a related culture and similar charms to those of Tahiti. It would be difficult to cover the vast literature on the subject, particularly on Hawaii, Samoa, and the Marquesas. Whatever reference of these other islands is made here is relevant to our more restricted focus on Tahiti, the island of flourishing love, and her surrounding islands.

A Tropical Arcadia

Native names for Tahiti illustrate the island's mystical, physical, or cultural characteristics: The Land of the Double Rainbow, Great Tahiti of the Golden Haze, Mounting Place of the Sun, The Gathering Place. For Europeans, it was seen as the voluptuous island, the innocent island.

This western interpretation, anchored in our vision of the South Seas, is of course not exclusively restricted to Tahiti. Other Polynesian cultures have acquired the same stereotypical characteristics. There is in my kitchen a magnet bought on the Big Island, figuring a Hawaiian hula dancer who seductively sways her hips and her grass skirt every time we open the refrigerator door. Obviously, the feminine languor and enticing dances we associate with Polynesia are not exclusive to Tahiti. In fact, most of us in North America have first experienced them in the Hawaiian Islands.

However, what happened in Tahiti is unusual. Rather than episodic information reaching Europe piecemeal over time, it took little more than a quarter of a century—between Wallis's first arrival in 1767 to Bligh's last voyage in 1793—to provide a strong and steady stream of extraordinarily similar and detailed reports. These came from British and French navigators and naturalists, surgeons and officers, reinforced by sailors' tales, who succeeded in evoking the picture of a Pacific paradise in their contemporaries' imagination. Their rank and education were often reflected in their style, but the tenor of the discourse was unvaried and nearly unanimous.

More than simply awakening a passing interest, their reports fitted with the great intellectual movements of the time. "The discovery of the Society Islands gave initial support to the belief that a kind of tropical Arcadia inhabited by men like Greek gods existed in the South Seas… The opening of the Pacific is… to be numbered among those factors contributing to the triumph of romanticism and science in the 19th century world of values," writes Bernard Smith in his study of the European vision of the South Pacific during that period (1960:1).

Whether these Europeans were personally seduced or disturbed, they limned the same picture of lush sustaining nature, enjoyment of life, lascivious dancing, and beautiful and easily accessible women. In editing for publication Cook's first voyage, John Hawkesworth had foreseen that many of the details would shock the British public and had suggested they do not stand in judgment, since morality was only little more than a matter of social custom. Worse details would come later, but initially, only licentiousness was deemed to be a possible bone of contention.

The first to arrive, Samuel Wallis on the frigate *Dolphin* and Phillip Carteret on the *Swallow,* had been sent to discover *Terra Australis* by the British admiralty. They had been charged with the task of collecting information about "the Genius, Temper, and Inclinations of the Inhabitants of such Islands or Lands as [they] may discover… but if no such case, to take Possession of such Land or Islands so discovered for His Majesty" (Hale,

1967:142). This was indeed very much the order of the day: explore, discover, describe, and, if not "civilized," annex. Wallis and Carteret did not discover Australia, but found Tahiti on the way, with its much needed fresh water and food, and something that part of the world would soon become famous for: its enchantingly free women.

One should probably pause here and wonder, as does Tahitian writer Chantal Spitz, what island women, used to well built and well bathed Tahitian men, found so attractive in these Europeans when they first arrived. They were hairy, unwashed, with bleeding gums and bad breath before the restorative qualities of Tahiti's climate, food, ocean, and sunshine could make them presentable. Lest we think Spitz biased, this is how W. Waldegrave compared "strong, muscular, and athletic" Tahitian men to English sailors in his 1833 *Journal of the Royal Society*: "When standing by the sailors, the natives looked large, their well turned muscles, erect carriage, and graceful walk, gave a very striking appearance" (Danielsson, 1956b:15). Could it be that exoticism works both ways? There is no doubt that the Europeans represented the unknown, even if their arrival had been foreseen in mysterious ways. They may have been seen originally as emissaries from the gods, the only explanation for their arrivals in such extraordinary vessels.

Initially, there was on the part of these women no explicit expectation of receiving payment for services rendered; prostitution had not yet been invented in Tahiti. However, it soon became obvious that the sailors were only too willing to trade gifts for the women's favours, and thus contribute to the local economy.

Wallis, unlike those who were to succeed him on the island, had a rather surly view of these attractive women. He witnessed, soon after the ship's arrival, their readiness to trade sex for modern iron tools, particularly the nails from his ship. Iron was not unknown to the islanders since many objects had been salvaged from one of the Dutchman Jacob Roggeren's three ships, wrecked on a nearby island a few decades earlier. With the prestigious red feathers, iron was the most sought after object of desire, and prized over any other trade goods.

Wallis and succeeding captains would have to watch constantly for their sailors' trading stolen nails to repay the native women. Those underhanded trades would not only endanger the safety of the ships, but also undercut the value of iron objects officially given to chiefs and other important personages to establish legitimate connections. Wallis's anger was thus justified, as John Hawkesworth explains. "While our people were on shore, several young women were permitted to cross the river, who, though not averse to the granting of personal favours, knew the value of them too well not to stipulate for a consideration: the price indeed was not great, yet it was such as our men were not always able to pay, and under temptation, they stole nails and others from the ship." The main problem was that these nails were already in use and, therefore, the sailors "drew several out of different parts of the vessel, particularly that fastened the cleats to the ship's side. This was productive of a double mischief: damage to the ship, and a considerable rise at market" (Hanbury-Tenison, 2005: 404).

George Robertson confirmed the extent of the pilfering on the *Dolphin*. "The carpenter came and told me every cleat in the ship was drawn, and all the nails carried off… The boatswain informed me that most of the hammock nails were drawn, and two thirds of the men obliged to lie on the deck for want of nails to hang their hammocks" (Kirk, 2011:48).

We will see later that Wallis would have another problem with the relations between his crew and Tahitian women. But this would not emerge until the arrival nine months later of the Frenchman Bougainville, who claimed Tahiti for his king and country, unaware that Wallis had already done so for his own king and country. The English would only discover that Bougainville had been present on the island when a native brought a French axe for repair to Cook's blacksmith.

Louis-Antoine de Bougainville, arriving so soon after Wallis, had been appointed by Louis XV of France in 1776 to circumnavigate the globe. He landed in Tahiti the following year and, utterly charmed with the beauty of the island, gave it the mythical name of New Cythera, after one of the Ionian islands reputed to be the site of Aphrodite emerging from the foamy sea. Aphrodite Cythera, the goddess of love: the evocation must have been spontaneous for Bougainville.

Educated men of his time knew Greek mythology well. Boticelli's *Birth of Venus* was as familiar to him as it was to his countrymen of like rank. Moreover, as a Frenchman, he would have been aware of Watteau's two *"fêtes galantes"* paintings depicting Cythère as the island of love:

L'embarquation pour Cythère of 1717, now in the Louvre, and *Pélerinage à l'Ile de Cythère* of 1721. The name perfectly fitted the mood.

He described this new Cythera as "an earthly paradise where men and women live happily in innocence, away from the corruption of civilization." He continued, "One would think himself in the Elysian fields." The allusion to Cythera and the Elysian fields was taken up again by the *Pandora*'s surgeon, George Hamilton, in his account of his voyage to Tahiti. "This may well be called the Cythera of the Southern hemisphere, not only from the beauty and elegance of the women, but their being so deeply versed in, and so passionately fond of the Eleusinian mysteries, and what poetic fiction has painted of Eden, or Arcadia, is here realized, where the earth without tillage produces both food and clothing, the trees loaded with the richest fruit, the carpet of nature spread with the most odoriferous flowers, and the fair ones over willing to fill your arms with love." (Gesner, 1998:37-38). Here, indeed, is the duality of the Tahitian myth perfectly voiced.

Bernard Smith suggests that Bougainville's description became permanently stamped upon the Europeans' imagination. "The country was so rich, the air so salubrious that people attained to old age without inconvenience. Indeed, the island was a healer. Men rotten with scurvy regained their strength after spending one night there" (Smith:1960:26).

How could Europeans have formed any other image of such a land than the one described by Bougainville: the Garden of Eden itself? How could they not have been thrilled with the existence of such a place that confirmed their dawning belief that nature provided the ideal source for the flourishing of human generosity and goodness, a reflection of the land itself and an expression of their nostalgia for man's loss of innocence? They read Bougainville's paean, "One gathers fruits from the first tree he meets with, or takes some in any house he enters.... I thought I was transported into the garden of Eden; we crossed a turf, covered with fine fruit trees, and intersected by little rivulets, which kept up a pleasant coolness in the air ... Everywhere we found hospitality, ease, innocent joy, and every appearance of happiness among them" (Smith, 1960:28-29). Smith concludes: "The land, in short, was like Paradise before the Fall of Man."

Paul Fussell writes that the "pastoral impulse—that is, the hankering after a simpler form of the world—is never destroyed. Founded as it is on the bedrock of universal human wishes, it merely changes its external shape from age to age, incarnating itself now in this form, now in that but never vanishing." He then quotes Renato Poggioli, the contemporary Italian critic who wrote in *The Oaten Flute,* "The poetry of the pastoral embraces both longing and wish fulfillment... The pastoral fallacy and its equivalents are deeply rooted in human nature; this explains the recurrence or permanence of their manifestations and the survival of pastoral make-believe even in such an Iron Age as ours" (Fussell, 1988:151-52). The eighteenth-century Europeans did not have to look far to find the embodiment of both their longing and their wish fulfillment in the reports they read about Tahiti.

However, Bougainville was soon faced with the problem that had plagued Wallis a few months earlier. He wrote, "The men... pressed us to choose a woman, and to come on shore with her, and their gestures denoted in what manner we should form an acquaintance with her. It was very difficult, amidst such a sight to keep at their work four hundred young French sailors." One may wonder which of the two, Wallis or Bougainville, had a more realistic understanding of what they actually witnessed. Where Wallis saw a nails-for-sex exchange, Bougainville, unstoppable in his enthusiastic description of idyllic island life, went on.

> I can tell you that it is the only place on earth where live men without vices, without prejudices, without needs, without disputes. Born under the most beautiful sky, nourished by the fruit of a fecund and uncultivated earth, ruled by the fathers of families rather than by kings, they know no other god than Love. Every day is consecrated to it, the whole island is its temple, all the women are its altars, all the men sacrifice to it. And how about the women, you will ask? The rivals of Georgian women in beauty and the sisters of the naked Graces (Salmond, 2009:19).

The naturalist Philibert Commerson, on Bougainville's expedition, confirmed that his captain "believed to have found in Tahiti Thomas More's Utopia... From the beginning his vision of Tahitians and their customs was tainted by this notion" (Lavondès, 1990:90). Commerson himself shared wholeheartedly Bougainville's vision. He is said to have held no reservations about what he saw as an admirable and balanced society. He saw no conflicts, only love, music, and games, and felt that whatever labours were required counted for little in an island so generously endowed by nature. He also believed that Tahitians, so harmoniously connected to their environment, could teach Occidentals a profitable lesson. Both men illustrate perfectly the current of thoughts prevalent among intellectuals at the end of their century.

It was generally agreed that the Tahitians had such nobility as to strike at the hearts of the visitors. Many natives offered special bonds of friendship, the *taio*, which gave the Europeans an insight into the Tahitians' social structure. John Marra, on the *Resolution*, wrote that they show "fidelity to those who condescended to place confidence in them as particular friends. To such, there is no service that they will not readily submit, nor any good office that they will not willingly perform; they will range the island through to procure them what they want, and when encouraged by kindness and some small present and tokens of esteem, no promises or rewards will influence them to break their attachments" (Marra, 1775:43).[7] Although there is certainly some innocence in the statement—for how can one make such definite assertions about a freshly encountered culture?—there is enough evidence that strong ties of friendship united some of the captains and naturalists with the important people on the island. All of them wrote about their special Tahitian friends and the protection they offered.In the face of Bougainville and Commerson's praise of Tahiti and her people, John Gilbert comments, "the French were more lyrical about the island and its inhabitants than the prosaic British. It was their extravagant praises that built in Europe the legend of a South Seas paradise where the 'noble savage' lived in a state of blissful innocence" (1973:280-81). However, it is possible that Bougainville's enthusiastic praise was not perhaps as wholehearted as reported, for, even on such a short acquaintance, he had formed some misgivings. According to Gilbert and to Salmond (2009:112), Bougainville had heard reports of tribal feuds, ritual murder, and killing of unwanted infants, and had started to suspect that all was not as idyllic as it had first seemed. The same thoughts were to occur to other visiting Europeans, and may have felt like flies in an ointment that seemed too beautiful to contain them.

The fact that the more lyrical, and certainly more risqué, French reports were published before the prosaic English ones probably tainted public perception. The first to have been to Tahiti, Wallis, was not the first one to make his discovery public. Bougainville's writings received much wider publicity in France than Wallis's had in England. In fact, his reports were translated into English, Dutch, and German before Wallis's own version was even circulated at home. The myth had started taking root before the reservations and the warnings were able to tamper the enthusiasm.

Naturally, Bougainville and Commerson were not the only ones to have commented approvingly on the local mores. For Joseph Banks, sailing with Cook on the *Endeavour* and expressing his views with less Gallic enthusiasm but with equal conviction, Tahiti represented the Golden Age or Arcadia. The people he saw there had all the beauty of Greek statues, worthy of the "chisel of a Phidias or the pencil of an Apelles." Enhancing this physical beauty, was the people's sexual freedom, which, wrote Bernard Smith, "filled him with admiration and delight, but in his English way he was more circumspect about it than the Frenchman [Bougainville]." In a letter intended to amuse the Princess of Orange by describing "the manners of Otaheite," Banks wrote that Tahitian women were "the most elegant in the world… Their clothes were natural and beautiful… so natural that nature had endowed them with beautiful bodies." For him, Tahiti was "the truest picture of an Arcadia of which we were going to be kings that the imagination can form" (Banks, 1896: 74).

Banks' admiration for the Polynesian women's beauty is echoed in many other contemporary testimonies, but is not restricted to the Tahitians. George Forster, one of Cook's companions, praised a Tongan woman in these terms: 'Her stature was graceful, and her form exquisitely proportioned. Her features were more regular than any I have seen in these isles, full of sweetness and the charms of youth. Her large dark eyes sparkle with fire, and her ebon curls floated on her neck" (Forster, 1777: Vol. II-183). The French explorer La Pérouse reports from Samoa, "The women, some of whom were extraordinarily pretty, offered us not only fruit and fowls, but also themselves… Their manner was gentle, soft and pleasing" (1797: Vol. II, p. 254).

The ascetic Spaniard Boenachea put his own religious slant on what he found in Tahiti and focused mostly on the venality of the natives, particularly the women. He is somewhat reminiscent of Wallis in his resentment of the local customs. He and his crew had different relations with the local people than the French and English before them, as he had forbidden his men to have sexual relations with Tahitian women, as well as following the same obligation himself. Their far more restrained behaviour had disappointed the natives. He wrote, "Whatever articles were obtained from us on board this Frigate were got for [the women]: the others begged with exceeding importunity in the names of the women for whatever we had, so much that they became a great nuisance to us. They tendered their women to use quite freely, and showed much surprise at our non-acceptance of such offers. The latter were also wont to make advances themselves, but with some show of coyness" (Salmond, 2009:254).

With this exception, the Europeans seemed to believe that their rapports with the natives were more or less cordial, once the first contact (sometimes accompanied by skirmishes) had been profitably made for both parties. They usually commented on the hospitality of the people and on their sexual mores that sometimes spread over a larger area than Polynesia, and, in spite of ethnic and cultural differences, included Micronesian and Melanesian islands. However, Bengt Danielsson (1956b: 13) believes that the three characteristics of erotic laxity, beauty, and hospitality that make Tahitians and some other Polynesians so attractive often occurred elsewhere but were seldom found together on other islands than Tahiti.

Tahitian women's beauty, which may have been surprising to Europeans reading and hearing about it, and perhaps thought to have been exaggerated through the men's long months at sea, has since been confirmed by Gauguin's paintings, and by the many films and documentaries now available on Polynesian islanders. Their features are usually regular and generally close enough to European ones to be attractive to them at first glance.

One of the least flattering, and perhaps more realistic, descriptions of Tahitian women comes from Bjarne Kroepelien, who evaluated them according to Greco-Roman ideals and did not think them beautiful with their big legs, large feet, rather fat figures, and, for some, large and open nostrils inherited from ancestors from the Tuamotus. On the other hand, he wrote, "There is in them something sometimes found in images from ancient Egypt. Their facial expression and their skin are much the same; they have long, smooth black hair, sensual lips, and they smile like a Sphinx who has nothing to hide; their eyes are always black, and their eyelashes incredibly long" (2009: 91-93). His personal canons for feminine beauty may have been influenced by his Scandinavian origins, but they did not prevent him from finding love in Tahiti.

While the focus of interest and description was the women, Cook has also commented on the handsomeness of the men and their graceful walk, and the beauty of the landscape. Similarly, Darwin (1860: Chapter XVIII), who spent ten days in Tahiti in November 1835, was much taken with the natives. "I was pleased with nothing so much as with the inhabitants," he wrote, describing them as "tall, broad-shouldered, athletic, and well-proportioned." He even compared very favourably their dark skin to the pallor of his countrymen.

Gauguin painted both men and women in the same light and much of our apprehension of the beauty of the island and its people derives from these paintings. But we are sometimes warned about the accuracy of the painter's art. Paintings of idyllic scenes may rely less on reality than on "the expectations of a society and on the vision of Eden; Gauguin does not paint Tahiti, but his Tahitian dream," writes the art critic Jean-François Staszak (2004:360). Both reality and dream interact, very likely, but even if Gauguin only painted his own dream and that of his contemporaries, what still remains is a representation of everyone's vision of Tahiti.

The travel relations that first reached England were often received with a grain of salt, or even downright suspicion and sarcasm. Banks, in particular, who may have been indiscreet in his descriptions of his feminine conquests and the carefree life of the Tahitians, was soon the butt of mockery, particularly from James Boswell and Samuel Johnson, which did not prevent the latter from wishing to make a prolonged visit to the island.

Nevertheless, and whatever public opinion may have first been in England, in the Royal Navy "Tahiti was legendary as a Paradise on earth, with its warm, scented nights, delicious food and glorious, beguiling women," writes Anne Salmond. Each voyage confirmed the seamen's opinion and anticipation as they sailed towards the enchanting island. Indeed, Alexander Home, a midshipman on *Discovery*, wrote that "almost Every man had a girl that Eat as much as himself and we lived in the Utmost Affluence. The Over flowing plenty the Ease in which men live and the Softness and Delightfulness of the Clime, the women are Extremely Handsome and fond of the Europeans, prodigiously insisting and Constantly importuning them to stay, and their Insinuations are Backed by the Courtesy of the chief and the admiration of the people in general. It is infinitely too much for sailors to withstand" (Salmond, 2009:452).

David Howarth comments that, "Sex became for Europeans the most astonishing and important thing about Tahiti, and the most discussed" (1983:48). Thus, we should not be surprised when Hamilton, the *Pandora*'s surgeon, remarked, "This I believe was the first time that an Englishman got up his anchor, at the remotest part of the globe, with a heavy heart, to go home to his own country" (Gesner, 1998:59).

In spite of some of the earlier reservations at home, the consistency of these descriptions contributed to the eventual creation in European minds of the myth of a life of almost naïve sensuality spent in a plentiful and generous nature, unencumbered by the rules and constraints of the already decadent society they faced in their own countries. The Prince de Nassau-Siegen, a handsome and elegant man who travelled with Bougainville as a passenger, had been offered upon landing the gift of a beautiful young woman. He demurred in front of the large interested audience gathered to see him perform ("some fifty Indians"), but later wrote in his journal, "Happy nation that does not know the odious names of shame and scandal. If wise people enacted ceremony to seed the earth, why should the reproduction of the finest species ever created not also be a public festival?" (Dunsmore, 2002:95).

Naturally, Tahiti was not the only island to provide such enchanting encounters for the sailors. Danielsson relates the visit to the Marquesas of Etienne Marchand, a French fur trader on his way to the Pacific Northwest in the 1780s. Previous visitors to the Marquesas had not made a favourable impression on the natives; Mendaña in 1595 had been involved in bloody fighting, and Cook in 1774 had hardly been welcome, the natives being unwilling to barter. Marchand, on the other hand, was well received. There was no misconstruing the natives' welcoming intentions, as they freely offered water. Moreover, "the men who surrounded the boats informed our sailors, by unequivocal signs, that the women were at their disposal." The women were made at once welcomed by the young sailors and negotiations began. Danielsson reports that, "as neither of the contracting parties resorted to dilatory or evasive tactic, they were not slow to descend into the tween-decks of the ship to conclude the treaty." When the women reappeared on deck, they carried small tokens of the sailors' appreciation:

"nails, mirrors, little knives, some coloured glass, ribbons… which they had obtained in exchange for the only commercial asset they had to dispose of." Marchand later described the women swimming around the ship "with the skill and agility of sharks."

Somewhat surprisingly, after describing what was essentially a sex-for-gifts transaction and the women swimming away afterwards, Marchand could not resist showing off his classical education and transposing the whole event into a scene every refined European would recognize as being appropriate to Tahiti and other South Seas islands. "It was a bringing to life of that charming picture of the birth of Venus, in which Boucher's brush depicts young Nereids playing on the waves around the shell which bears the goddess. And what effect might not the arts of these sirens have on a young sailor who was no Ulysses!" (Danielsson, 1957: 42). It would be difficult to find a better example of the taste for embellishing what would otherwise be a plain retelling of an ordinary exchange between sailors and local women. Perhaps Marchand had felt the description might otherwise be seen as sordid, and believed that throwing in a few classical allusions would made it quite acceptable to the more cultured Europeans readers, who would then see the encounter as romantic and exotic, and more conform to their expectations. Yet, it is likely that these lovely Nereids playing in the waves were not actually prostituting themselves; in spite or perhaps because of the people's very free sexual mores, prostitution did not exist at the time. Naturally, when the women realized that what came naturally to them could be bartered, they saw no reason to deprive themselves and their people of what the sailors dispensed so readily. Neither would the men in their family later deem it inappropriate to act as go-betweens and negotiators. Little did these young women know that they and their people would actually be the ones to pay the greater price for these encounters with European sailors in venereal infections, infertility, and stillbirths for what seemed at first to be such a natural exchange.

Sailors, often young, unsophisticated men, sometimes liked to delude themselves into thinking that this exchange was totally consensual and even welcome by the women. Cook's surgeon, David Samwell, told of "exceedingly beautiful" women who "used all their arts to entice our people.. [and] they absolutely would not take no denial." William Ellis agreed, "There are no people in the world who indulge themselves more in their sexual appetites than these [young women]" some of whom "could not be more than ten years old." Both are quoted in Igler (2013:49-51), who comments that while contemporary reports often focussed on the willingness of the women, fulfilling most young men's desire for consensual sex, they usually ignored the nature of relative power in these sexual exchanges.

Igler also quotes Marshall Sahlins, writing that in Hawaii "sex was everything: rank, power, wealth, land, and the security of all these," a statement that would also apply to the similar Tahitian culture. Igler guesses at a system relying on the young women's "expectation of social advancement: increased mana (or spiritual power) taken from a stranger, [and] performing one's duty under the watchful eyes of village elders." There is little doubt that much collective good ensued from the originally free granting of these favours. At first, nails and trinkets were gratefully offered, but these gratuities soon led to an overt system of prostitution throughout Polynesia, with male relatives often serving as official pimps. On Cook's second voyage, George Foster described how some Maori men procured women for the sailors. These "often were the victims of brutality dragged by the fathers into the dark recesses of the ship, and there left to the beastly appetite of their paramours" (Igler, 2013:51).

Thus, what we have so far seen as a fair trade between willing women and eager seamen is far more complicated than appeared at first glance, as we mostly based our understanding on reports made by individuals with little grasp of Polynesian culture – who, moreover, would have much preferred to believe in these young women's natural inclination for freely dispensing their welcome.

The Noble Savage

In the intellectual climate of the times, England and France tended to be favourable to the notion of a simpler and more innocent life. More to the point, the interest was in discovering and defining the essential nature of man in its pure state.

The Earl of Shaftesbury had proposed in his *Inquiry Concerning Virtue* (1699) that a proper sense of morality could be obtained through one's spontaneous emotions and that mankind's essence was to be good. This work was later to become the source of the French Encyclopedist Denis

Diderot's *Essai sur le mérite et la vertu* (1745) and reach a greater European audience. However, it had been left to John Dryden to use the term "noble savage" for the first time in his play, "The Conquest of Granada" (1672), and to expound on the notion of "Nature's gentleman," uncorrupted by civilization. Another early contributor to the creation of the noble savage prototype was the French Baron de Lahontan, who had extensively visited North America and had written a widely read travelogue in which his experiences among the Hurons served to illustrate the freedom of their ways and the rationality of their conduct.

By the following century, a current of sentimentalism in philosophical and literary writings on both sides of the Channel had taken up the notion of the inherent goodness of man. In England, Jonathan Swift had described the perfect realm in 1726. "Friendship and Benevolence are the two principal Virtues among the Houyhnhnms; and these were not confined to particular Objects, but universal to the whole Race. For, a Stranger from the remotest Parts is equally treated with the nearest Neighbour, and where-ever he goes, looks upon himself at home" (1996:202).

This description of friendliness, generosity, and hospitality would have been recognized by the Europeans when they visited Tahiti half a century later. However, the major difference between the imaginary Houyhnhnms and the idealized Tahitians was less a result of belonging to two different species than in their respective attitudes towards conjugal duties. The former's society was exclusively ruled by reason and their attitude towards love, marriage, and fidelity was so exclusive that the sexual feasts afforded to the visiting sailors to Tahiti would have been incomprehensible to them. Although, there again, one should not generalize the sexual freedom of some to all, as unfaithful Tahitian wives were known to have been chastised.

The Tahitians certainly rejoin the Houyhnhnms in their idyllic manner of conducting their lives. Bligh himself writes that "they are kind and humane to one the other beyond a doubt and are tender parents. What then have we need of more to prove that in their natural civilization they have the leading virtues of a happy life" (Salmond, 2011:201). Bougainville and Commerson also marvelled at the "admirable balanced society without apparent conflicts" that epitomized for them the utopian side of Tahitian life. No doubt inspired by the influential Jean-Jacques Rousseau, Commerson wrote to the astronomer Lalande that in New Cythera could be found "the state of natural man, born essentially good, free from all preconceptions, and following, without suspicion and without remorse, the gentle impulse of an instinct that is always sure because it has not degenerated into reason" (Dunsmore, 2002: 94).

Much of the desire to discover how the "natural man" behaved was behind the observations conducted by the early navigators and naturalists. Certainly, they were interested in the fauna, flora, and geography, and so on, and they were most successful in mapping these out, but they also felt a strong curiosity towards those men they were discovering. "I had long wished to see a people following the dictates of nature without being bias'd by education or corrupt'd by intercourse with more polished nations," William Anderson wrote in his *Journal,* the ship surgeon sailing with Cook on his third voyage.

Others had different purposes for studying individuals untainted by civilization. Writing about the "Noble Savage" was also a pretext to criticize current society, particularly in France during the years leading to the Revolution. Many thought idealistically that a return to a simpler way of life could solve some of the problems of an unfair social structure.

Spearheading the movement was Jean-Jacques Rousseau, a friend of philosophers Diderot and David Hume, and a critic of Thomas Hobbes's belief that man living in a state of nature was evil because he did not know virtue. In 1754, Rousseau's position was that nothing was so gentle than man "restrained by natural pity from harming anyone" when in "nascent society"—that is, when placed by nature midway between "the stupidity of brutes and the disastrous enlightenment of civil man" (Rousseau, 1984:98). Thus, he believed that the child was essentially good but became corrupted by society and the influence of civilization, and wrote *Emile, or on Education* (1762) to test this belief.

Rousseau's philosophical constructs were not based on any real experience of the savage life he praised. This did not prevent Claude Lévi-Strauss, ethnographer and great admirer of Rousseau and Chateaubriand, to go beyond this apparent weakness and see Rousseau's attempt in the *Confessions* to "rediscover the union of the sensitive and the intelligible" as the first step in awakening consciousness and in developing a primordial "alliance against a society hostile to man." Lévi-Strauss concludes his essay on "Jean-Jacques Rousseau, Founder of the Sciences of Man" by showing how Rousseau teaches man "how to elude the unbearable contradictions of civilized life. For if it is true that nature has rejected man and that society persists in oppressing him, man can at least reverse the poles of the dilemma to his benefit and seek the society of nature to meditate there on the nature of society" (1976:40).

Voltaire, who stood for most forms of freedom, particularly that of thought, wrote his famous *Candide, ou l'Optimiste* (1759, translated into English in 1772), in which the simple joys of "cultivating one's garden" were opposed to the duplicity and violence of a complicated society. Voltaire had read Champlain's writings and the *Jesuits Relations*, which explains in part that the Enlightenment concept of the Noble Savage or the Natural Man did not come from Polynesia—although the harmonious beauty of the island and its inhabitants could have warranted such an origin—but from the colonies of North America, discovered earlier. The North American native was deemed to behave with honour and courage, and his life, grounded in nature, was admired for its apparent simplicity—a view encouraged by the Baron de Lahotan's reports on his stay among these natives. Similarly, Marc Lescarbot admires in his *Histoire de la Nouvelle France* (1609) the purity of these "savages," deemed to be nobly human and satisfied to live from what God had provided. This vision was not unanimous among those who had witnessed them at war. Bougainville, in particular, who had been appointed in 1756 as aide-de-camp to Montcalm when the war broke out between Britain and France, and who had fought alongside Iroquois warriors, did not think much of the idealistic notion fostered in Europe about the noble savage's actions in North America.

The Natural Man became the theme of several books. Bernardin de Saint Pierre's 1773 *Voyage à l'Ile de France was* followed in 1788 by the much acclaimed novel *Paul et Virginie*, based on his travel to Mauritius. As well as denouncing the horrors of slavery on Mauritius, where sugar cane was grown, he exalted the idyllic vision of a harmonious life.

The problem was that, for some purists, it was difficult to judge what constituted a state of pure nature. Salmond (2009:399) refers us to a conversation between James Boswell and Samuel Johnson. The former had hoped to join Cook's next expedition, wanting to spend three years either in Otaheite or in New Zealand "in order to obtain a full acquaintance with people so totally different from all that we have ever known and be satisfied what pure nature can do for man." He was to be deterred by Samuel Johnson, whose argument was as follows: "What could you learn, Sir? What can savages tell, but what they themselves have seen? Of the past, or the invisible, they can tell nothing. The inhabitants of Otaheite and New Zealand are not in a state of pure nature; for it is plain, they broke off from some other people. Had they grown out of the ground, you might have judged of a state of pure nature" (Boswell, 1962: 559-60).

Nonetheless, when Mai or Omai (or even Omiah), the young Polynesian brought by Captain Furneaux to England, made the social rounds, he much impressed London society by his "natural" ways, as well as his stature, appearance, and tattoos. Featured in Robert Dodsley's influential *Annual Register*, he was even one of the *Characters* for the year 1774, a social consecration to which many Englishmen aspired.

The philosophical notion of the Natural Man became somewhat diluted with Fanny Burney, the popular English novelist, diarist, playwright, and sister of James Burney, one of Cook's companions. Salmond (2003:297-98) reports Burney's description of Mai's foray into the salons of the day. Comparing him to Lord Chesterfield's son, who for all his education was "a meer *pedantic* Booby," she writes, "I think this shews how much *Nature* can do without *art*, than art with all her refinement, unassisted by *Nature*," apparently considering that the naturally good manners of the young Tahitian could be equated with the other virtues the Noble Savage was deemed to possess.

George Hamilton, the *Pandora*'s surgeon, waxed philosophical when he compared what he saw in Tahiti in 1791 with the prevailing European thinking of his century. "[Tahiti] affords a happy instance of contradicting an opinion propagated by philosophers of a less bountiful soil, who maintains that every virtuous or charitable act a man commits is from selfish or interested views. Here human nature appears in more amiable colours, and the soul of man, free from the gripping hand of want, acts with a liberality and bounty that does honour to his God" (Thomas, 1915:108-09). His views are perhaps somewhat surprising in that the *Pandora* had been sent to hunt down those who had defected to the very life he so admired, the *Bounty*'s mutineers.

There is no doubt that the discovery of Tahiti and the many journals, reports, or letters written about it resulted in strong literary reactions and artistic depictions in Europe. Both literature and paintings often included many references and allusions to Greek mythology, so much resemblance did artists and writers see between the idealized version of life in Tahiti and the physical perfection of Greek gods and heroes.

Bougainville's final depiction of Tahiti, in fact his farewell to the island, epitomizes the general feeling of admiration felt for the island and the islanders, a sentiment he would be the first and most eloquent to publicize in Europe. "Nature had placed the island in the most perfect climate in the world, had embellished it with every pleasant prospect, had endowed it will all its riches, and filled it with large, strong, and beautiful people… farewell, happy and wise people, remain always as you are now. I will always remember you with delight as long as I live I will celebrate the happy island of Cythera: it is the true Utopia" (Salmond, 2003:53).

And celebrate it he did. However, one should remember that Bougainville stayed only nine days in Tahiti and that his vision was strongly influenced by his classical education. He was not to face, at least directly, any of the unpleasant consequences of Tahiti's contact with Europeans, and as he left the island he was almost himself in a state of innocence.

His words find a strong echo in Commerson's own farewell to Tahiti. "One must call it Utopia, both in fact and in name, as indeed it was founded by Themis, Astrea and Venus, and freedom, the most precious of all gifts, far from the vice and dissentions of other mortals, an eternal and sacred state where unbroken peace reigns among the inhabitants as in a very holy Philadelphia." He particularly admired the patriarchal government "by which are given to outsiders (even ungrateful ones) the most complete and trusting hospitality and the free and generous gift of every kind of riches the earth provides" (Gunsmore, 2002: 90).

The French may be forgiven for their vision, untainted by the reality of tribal warfare and rigid hierarchy that prevailed in Tahitian society. They only learned through hearsay of some aspects of the darker side of Tahitian culture and preferred to continue believing in an idyllic way of life. Unfortunately for Tahiti, their vision prevailed and is the one, fabulously adorned with all its classical references, that first circulated in Europe and dominated European imagination.

It would take some time for others to report on what the same Europeans would have considered serious flaws in the culture. These could not be reconciled with the notion of the noble savage, his inherent goodness of heart, and his admirable life. How could they accommodate themselves with what they found there, including cannibalism (although more prevalent in the Marquesas than in Tahiti) and infanticide (mostly practiced in a certain class of society, the *ariori),* while still proclaiming the nobility of the natural life? They knew about these, and had no explanation for them other than that Tahitian culture permitted them. They were wise enough to realize that their ethnocentric observations could not penetrate the society they were discovering, with both its worthy traits and its frightening deviations. In other words, they understood they were facing a culture as complex as their own, one they had no power to change to suit their own beliefs. This task would be left to the missionaries.

Going Native

The beauty of the Polynesian islands, the welcoming manners of their inhabitants, the sexual encounters in which they so easily engaged, all were to attract and seduce Westerners throughout the eighteenth, nineteenth, and twentieth centuries, with the appeal of the myth and the cliché remaining for some as strong as ever. The return to a presumably more innocent and simpler way of life is an enduring myth, to which modern people are not immune. Where else to find it but in Tahiti and nearby islands? We need only think of Melville's stories, Gauguin's paintings, or Jacques Brel's songs.[8] This nostalgia for the simplicity of a younger and unsophisticated world is one of the strongest urges that led Europeans to live, at least for a while, as Tahitians, pretending to "go native" in so doing. Among the first to adopt a few of the local practices (at least those that were not so overtly lax as to cause concern to their captain) were some of Bligh's men. He allowed the crew to live on shore and encouraged them to care for the breadfruit plants while these were getting ready for transplanting. It took little time for the men to become superficially socialized into the Tahitian culture and adopt some of the local customs. Many sailors, as well as "forming connections with" the women, as Bligh put it, and eating the local food, also chose to be tattooed in the local manner. Fletcher Christian and George Stewart of the *Bounty* were decorated with a star on their chest. When the *Resolution* briefly visited in 1777, many were tattooed on their arms, including Captain Sever, the surgeon Bowes Smyth, and particularly lieutenant Watt, who displayed tattoos almost over all his body.

The *Stratford*'s surgeon, William Dalton, on his second visit to the Marquesas in 1824, described the process he underwent: "The constant hammering at the skin, or into it, with considerable violence, irritates the whole frame, and the constant wiping off the blood with the *tappa* [bark cloth] is worse. However, as the work proceeds, the flesh swells up, which gradually benumbs the part." Having borne the ordeal for four hours the first day and three hours the second, he was rubbed with coconut oil and made to rest in the shade of a tree, feeling "a little faintish." A few days later, "the swelling all went down, the outer skin peeled off," and, while unrecognizable (his face had not been marked but his whole body was tattooed), he had completely recovered (Druett, 2001:175-76).

The tradition of tattooing in the British Navy can probably be traced to these early experiences in Polynesia. By 1802, there were places in English ports where Royal Navy sailors could be tattooed—often with a crucifix on their backs, in the hope that such an emblem would prevent them from being flogged, or at least that they would be flogged with less vigour (Kirk, 2011:143).

A certain Julien Viaud was introduced to tales of Polynesia by his brother Gustave, a French navy officer. Julien was immediately seduced when, having also become a sailor, he was stationed in Papeete for two months in 1872. He lived among local people, imitating their customs, trying to learn their language, wearing their clothes, and even adopting for his literary career the name they gave him: Loti. He was infatuated with the Tahitian way of life. "In Oceania toil is a thing unknown. The forest produces spontaneously everything necessary for the nourishment of these carefree tribes; the fruit of the bread-tree and wild bananas grow for everybody and suffice for all. The years slip by for the Tahitians" (1991:40).

The book he wrote as Pierre Loti in 1880, following an idyll with Rarahu, a young Tahitian woman (or, more accurately, a composite figure of the various young women he met there) became an immediate success in France. I will later expand on this popularity and the part he played in perpetuating the Tahitian myth in France in the 1880s.

It was an understandable temptation for Europeans to adopt local customs and life tempo, since those were the very lures that drew them to Polynesia. Danielsson encountered many such individuals while living on Tahiti and Raroia. Most were ordinary white men whose only desire was to end up blending among the rest of the population. A few were well known, such as the English poet Rupert Brooke, called Pupure ("Fair") by the Tahitians, who spent several discreet months wearing the comfortable local pareu and seldom going into town, while writing some of his best poems.

Gauguin epitomizes the white man's search for a natural Eden. His road before finally getting to Polynesia in 1891 had been long and arduous, but there he remained and there he died in 1903. He is buried in the Atunoa cemetery on Hiva-Oa, the island where he last lived, only a few feet away from Jacques Brel, another artist seduced by another mirage, that of the sea.

Gauguin had intended from the start to live according to local customs; the houses he built were strongly influenced by those of the natives, and the young women he took as wives were local *vahines*. In *Noa Noa,* he describes one of them, Teha'amana, as a "beautiful, golden-yellow flower, full of fragrant Tahitian *noa noa*, and I worshipped her as an artist and as a man." Yet, Teha'amana, very likely following the custom and probably true to form, had "many lovers, whom she used to meet in the thickets… when Gauguin was under the impression that she was gathering food or gossiping with friends," reported to Danielsson (1965:121), who had been told about it by two women who belonged to Teha'amana's group. Surely, had Gauguin known of her trysts, he could hardly have complained since such freedom was a part of the very culture that had attracted him there. He was also well aware of the differences in the relation between love and sexual congress in his native and his adopted land when he wrote the following aphorism, "In Europe men and women have intercourse because they love each other. In the South Seas they love each other because they have had intercourse." He also commented, very unpleasantly, "Women here don't beat around the bush… An orange and a glance are enough. Oranges cost between a franc and two francs, so there is really no reason to save on them" (Danielsson, 1965:129). At least Wallis's men had traded nails, which were no doubt of much greater value to the community than Gauguin's cheap oranges.

Yet, his Tahitian exile was still a quest for the natural man. In 1886, he wrote to his long-suffering wife Mette, "Christianity and civilization have tried to abolish man's belief in himself and in the beauty of primitive instincts, with the result that it has become a myth, but a myth is still alive within every man. I want to make the myth a reality again" (Danielsson, 1965:136).

Sadly, those nineteenth-century attempts at reliving the early Tahitian culture, the one described upon contact with the Europeans, took place on an island that had already almost lost the soul that animated it a century earlier.

[III]
THE VAHINE

AT THE HEART OF EUROPEANS' love affair with Tahiti is the *vahine*. The eighteenth-century reports clearly showed that, even with some serious misgivings and a great deal of misunderstanding, the local women's warm reception was much appreciated by the European sailors. There were officially no women among the Europeans who visited Tahiti, so the reverse did not apply at that time. We will have to wait until contemporary times to wonder whether *tane,* the *Ma'ohi* man, is as appealing to European women as *vahine* was, and still is, to European men.

How not to fall in love with these flower women, playing in the sea like dolphins? It is a mixed imagery, yet one that satisfies my sense of what they must have looked like. Many had commented how cheerfully they would swim out to meet the ships, diving in and out the waves. What an extraordinary and irresistible contrast they must have made with European women! They were also easy to please, their demands few beyond the much-craved iron nails. Melville, after describing his lovely Fayaway as a "child of nature, breathing from infancy an atmosphere of perpetual summer, and nurtured by the simple fruit of the earth, enjoying a perfect freedom from care and anxiety," further explains how she and her friends made themselves even more beautiful by simply adorning themselves with flowers. "Flora was their jeweller. Sometimes they wore necklaces of small carnation flowers, strung like rubies upon a fibre of tappa, or displayed in their ears a single white bud... looking like a drop of the purest pearl... The maidens of the island were passionately fond of flowers; a lovely trait in their character (1982a:108-09)." They seemed such child-like creatures that it must have been difficult to accept their venal side, almost entirely created by the white man's own desires and, eventually, by the brown man's own greed.

In the nineteenth and twentieth centuries, works of fiction continued to nurture the image of the Tahitian woman in European and American imagination. She has been illustrated in films, books, songs, and paintings, and usually depicted as being either naïve or venal, but always seductive. Let us consider a few renditions of the *vahine* through different media, where stereotypes flourish unabashed.

Tahiti Through the Silver Screen

The navigators expounded on the Tahitian way of life as they perceived it, and it is thanks to their writings that the Arcadian myth has been so firmly implanted in modern imagination. However, Bligh had an even more significant impact. His misadventures on the *Bounty* so struck both writers and filmmakers that he has become emblematic of those who sailed to Tahiti and trod on New Cythera.

Some of us, who may not be familiar with Cook, Banks, or Bougainville's writings, have perhaps read Pierre Loti, Herman Melville, Robert Louis Stevenson, or James Michener on the South Seas. Certainly, no one has reached a wider and more universal audience than Gauguin with his evocative Tahitian paintings. However, their role in spreading the image of Tahiti is more than matched by that of the American film industry. It would be hard to determine who—of the writers and painters on the one hand and of Hollywood on the other—has played a larger part in feeding the island myth in indelible pictures to the greater number of people.

Being curious about our persistent conception of island life, I asked some twenty people, none of whom had visited French Polynesia, what the name "Tahiti" evoked for them. They responded unanimously by mentioning (in no particular order) Gauguin's paintings, the *Mutiny on the Bounty*, and an image applicable to any tropical island with golden sands and swaying palm trees. Among the older ones, two also remembered Dorothy Lamour and her sarongs.

The film that had the strongest impact in defining in popular imagination the innocent yet licentious Polynesian lifestyle is *Mutiny on the Bounty*. Of the three main versions, I have chosen as a reference the 1962

Metro-Goldwyn-Mayer production featuring Marlon Brandon as Fletcher Christian and Trevor Howard as Captain Bligh. It evoked the terrestrial paradise we know and did not reveal anything the audience was not already prepared for. The pristine Garden of Eden image was already well anchored in people's mind through a previous Hollywood version of the same story. We find it again in *Captain Horatio Hornblower*, another popular motion picture produced twelve years earlier. Only a few minutes into the film, a sailor is heard complaining that instead of being stuck in Hornblower's becalmed ship, with cargo and destination unknown, they could be with "brown-skinned girls, bread growing out of trees, where the *Bounty* went." Everyone in the audience would have recognized the reference.

In the film, Bligh explains his mission to bring back to England the breadfruit trees to feed the slaves in Jamaica. The critical unknown of the endeavour is to determine the beginning of the five-month dormancy period of the trees, during which they cannot be transplanted. Hence the horrendous passing of the Cape of Good Hope in the winter with Bligh hell-bent on arriving as soon as possible, before the trees become dormant. The film does not spare us the evidence of the violent but shorter passage of the Cape, as time is of the essence, nor the brutality of the captain, so the point is well made. But, not satisfied with an ordinary island with a good climate and sufficient food, which would have been reward enough, they land in Tahiti.

Bligh has already warned the crew. "Those savages have absolutely no conception of ordinary morality and you will no doubt take full advantage of their ignorance. It is a matter of supernatural indifference to me whether you contaminate the natives or whether the natives contaminate you." His only interest is the mission with which he has been entrusted, and he warns his crew of the dire reprisals that would be meted out to any man who endangered it.

The first apparition of the natives shows some well built young men, their heads circled with foliage, a *pareu* draped around their waist, and carrying bunches of bananas. They notice the large ship in the bay and blow a seashell to alert the village. Men rush to their canoes and seemingly within minutes, the *Bounty* is surrounded with outriggers to the sound of drumming. They escort Bligh, Fletcher, and a few sailors in the ship's boat as it heads for shore. The Englishmen are immediately surrounded by young and beautiful people, all wearing foliage around their heads, the women's naked breasts covered with flower garlands and their long flowing dark hair. With the exception of the elderly king and the middle-aged translator, all the inhabitants are young and beautiful.

The royal barge arrives and trinkets brought by Bligh are received with childish glee by the king against the agreement to provide as many breadfruit plants as the Englishmen will need. Unfortunately, to Bligh's barely hidden fury, the plants are just entering their dormant period and the crew will have to stay there and wait for several months.

Delightful scenes are shown on the screen, depicting women and men fishing in the lagoon. The women line up in a flowery and charming row across the lagoon ("like a string of live pearls," remarks an enchanted sailor), while the men in their advancing canoes flog the water to scare the fish into the arms of the waiting women. The circle of women gets smaller and smaller until all the fish are herded into a reduced space and can be speared or caught by hand. The English sailors join in and everybody has a marvellous time, for the natives are always smiling and good-natured. The fish will be eaten at the feast given in honour of Bligh and Christian.

There is only one exception to the natives' good humour, one that serves to confirm Bligh's understanding that they have "no conception of ordinary morality." During the feast, women and men are seen dancing in a manner that polite English society of the time would have found exotically and abhorrently lascivious. With brown hips gyrating and bellies thrust forward, the dances are lewd enough to seduce the appreciative sailors, and the sexual promiscuity is instantly manifest. Christian has caught the eye of the king's daughter, Maimiti, who is only too willing to entice him into the bushes, when Bligh interferes and sends Christian back to the ship. The king, irate and humiliated, threatens, "If king's daughter not good enough for England, breadfruit not good enough for England." Bligh deals with the awkward situation by ordering Christian to return to shore. He fumes that "in a civilized society certain lewd intentions towards female members of one's family are an insult. But in Tahiti, the insult lies in the omission of these lewd intentions. You have offended his code of etiquette… Make love to that dammed daughter of his!"

The voice-over gives tongue-in-cheek cultural insights into this amorous behaviour and how it found an echo in the crew. "The ship's company were pleased to learn that Tahitians considered love-making a gesture of good will. It may sound improbable, but the good will in the hearts of our crew turned out to be practically boundless. The Tahitian ladies gave them every opportunity to prove it. You see, to the Tahitians, a light skin was a mark of beauty. Men with fair complexion, like Englishmen, were regarded as beautiful. No matter how nonsensically their features were arranged, they were beautiful and they could do no wrong. So, they did no wrong at every possible opportunity." To illustrate the lesson, the screen shows inviting gestures made by flowered and undulating women, and couples draped in garlands strolling hand in hand into the sunset. "A man can find a life for himself here," muses one of the future mutineers.

The final evocative touch is seeing Christian and Maimiti kissing in the surf, reminiscent of the famous passionate embrace between Burt Lancaster and Deborah Kerr in *From Here to Eternity*, shot only a few years earlier, and which movie goers of the day would immediately have recognized as a symbol of unbridled sensuality and passion.

When Marlon Brando married the young Tahitian actress Tarita Teriipaia (Maimiti in the film) and bought a small atoll to build his home, it only seemed to confirm the verisimilitude to the fictional love story, merging and confusing fantasy and reality. Tahiti was definitely the island of love.

It would not be the first, and certainly not the last time, that the South Pacific islands figured in films and television shows. Robert Schmitt has tracked down 165 movies that had a Polynesian locale, out of the 21,000 made between 1913 and 1943. Usually with little artistic merit and ignored by serious critics, they nevertheless served to feed popular imagination with stereotypes that would later serve to promote post-World War II tourism to these islands. While Hawaii, understandably dominated the scene, eleven of these movies focused on Tahiti. Most were shot in the Hollywood studios, but some were filmed in Hawaii (27) and Tahiti (5), and twelve others were shot in exotic sites, such as the Bahamas, Cuba, Florida, Puerto Rico, or Catalina Island, substituting for the South Seas. Similarly, nowadays, any dark-haired woman with perky breasts and golden skin, will substitute for a *vahine* on postcards flaunting "Tahitian beauties."

Some of these earlier movies featured violent natural occurrences, such as hurricanes, typhoons, crocodiles, sharks, and giant octopi. Volcanoes erupted in at least twelve films, and ten others had the actors facing high winds and torrential rains. The islanders were mostly seen "battling typhoons and lava flows" in their daily activities, "swimming and diving for pearls, [being] attacked by sharks, swordfish, giant rays, octopi, and clams, all of which attain prodigious sizes in South Seas movies" (Schmitt, 1968:434-450).

These old movies were sometimes shown in the places they were deemed to depict. James Michener describes in his chapter "Polynesia" the reactions they evoked from the audience. According to a Papeete cinema owner, the favourite picture, shown repeatedly by popular demand in Tahiti in the 1940s, was *Aloma of the South Seas.* It is "the purest corn ever grown, and in Technicolor. It is so completely ridiculous that it packs them in, howling with delight at the blunders. Another sure-fire hit is *South of Tahiti*, which has a big scene with a tiger. Says a critic, 'The house goes wild when that beast appears, because on most islands here there aren't even any mice'" (Michener, 1951:57-58).

In these films, the familiar and attractive indolence of island life has gone. Rather, they bring forth yet another myth: the danger of the unknown, amply illustrated in the natural forces and fictitious savage beasts faced by the Hollywood movie heroes. Closer to our own familiar stereotype implanted since the early navigators, we also find interracial romances, with an emphasis on "the native or half-native heroine… at the expense of the native male," and "the white interloper as lover of the beautiful native princess," as J.C. Furnas, another film critic, put it. He added, "The South Seas legend is most easily illustrated by movies. It consists of sex, bare skin, idyllic settings, shoals of girls, few men except the male lead, and is strung on a plot as foolish as it is stereotyped… Casts and quality of photography change—Hollywood's South Seas do not" (Schmitt, 1968:435).

We note that modern Tahitian literature has also considered this reversal of the traditional roles, at least in Chantal Spitz's *Island of Shattered Dreams.* Schmitt reports that four of the twenty-one movies on interethnic romance "reversed the usual roles and presented *haole* [white, in Hawaii] girls involved with island men. All but one of the men were treated kindly by the moviemaker. Interracial triangles were not uncommon." In the latter, *haole* girls were sometimes seen torn between a *haole* and a half-caste man.

Actual South Seas locations were more often used later than in the past, and several movies were filmed on site, notably the three versions of *Mutiny of the Bounty* (1935, 1962, and 1984), filmed in Tahiti, Bora Bora, and Moorea, forever imprinting the beauty of these islands in the viewers' eye and mind. More generally, within the last thirty years the South Seas served as background to a number of films and television shows. *Blue Lagoon* (1980), *Return to Blue Lagoon* (1991), and the recent Tom Hanks' *Castaway* were filmed in one of the Fijian islands, while *Couples Retreat* (2009) was filmed in Bora Bora. The *Bachelorette,* a televised reality show, had its 2010 finale shot in Taha'a and Bora Bora, and another, *Survivor* (2000-2010), variously took place in Borneo, Vanuatu, Samoa, Micronesia, Palau, Cook islands, and the Marquesas. The latter event, which occurred in 2002, is still mentioned locally with pride, and figures in touristic pamphlets, as if the Marquesas' very real charms needed this fictional validation.

The point is that, whatever their topographies, their histories, their idioms, their cultures, or the physical appearance of their inhabitants, all South Seas islands are presented to the general public as being interchangeable. All equally evoke a leisurely life, a plentiful nature, physical beauty, and carefree indulgence, as well as all sorts of hidden dangers for the sake of the plot. On the whole, they are meant to illustrate Baudelaire's *"Invitation au Voyage"* to a place where all is *"ordre et beauté, luxe, calme, et volupté."*

In the Eye of the Beholder

The early Tahitians' culture was essentially oral and their artists aspired to become epic poets, renowned genealogists, and talented speakers. Their pictorial tradition was mostly expressed through perishable material and exposed to nature, including wood and stone (petroglyphs and *tiki*), and skin (tattoos).

Tiki (Nuku- Hiva)

It is mostly thanks to the various European artists sailing with the navigators that we can picture the Tahitians' scenery and the objects that constituted their daily life, their dwellings, their rituals, or their clothing at the end of the eighteenth century. Cook, for one, brought along three artists on his voyages: Sydney Parkinson, William Hodges, and John Webber. Commerson, a botanist sailing with Bougainville, left a profusion of sketches of natural species and artifacts. A few decades later, Dumont d'Urville also brought along painters on his visit to Polynesia in 1838.

John Webber, among the many sketches he made of the island, showed women involved in various performances. For instance, he drew two ladies wearing the ornate costumes of their rank, including half-shell coconut brassieres, bulky skirts, and a sort of rigid pleated peplum raised behind their backs. They are shown performing a stately dance, with two male dancers, knees wide apart, dancing beside them. In the background, four men play the drums. Palm trees and some vegetation dominate the background. The picture appears staged; the women are graceful and have a modest comportment with nothing in their movements to inflame men's imagination and desire. Actually, neither do they look particularly Tahitian.

Ladies Performing a Dance (John Webber)

Given their responsibility of informing on all aspects of Tahitian life, it is not surprising that eighteenth century artists were, on the whole, satisfied with including women as part of the general picture. They showed Tahiti's many attractions, but the women were only one component of the island's appeal. Artists mostly sought to represent them in a natural context and do justice to the appealing quality of their occupations. Under the classic influence of their art, they tended to develop a style sometimes known as the "South Seas mirage" (O'Reily).

The perspective changed with Paul Gauguin, the most famous painter of all things Tahitian, particularly women, whom he painted at every age, in every state of dress and undress, in every position, and most often in vibrant colours. "Post-modernist exoticism" is sometimes used to describe his style, which fixed forever the image of the *vahine*. After all the descriptions of Tahitian women limned in the eighteenth century, it really took Gauguin to display to Europe the faces and bodies of those exotic women already known to be enchanting. That Gauguin loved Tahitian women (or, more accurately, young Tahitian girls barely into their teens) is well known, and he painted them more or less true to life, as they were can occasionally be seen today in the odd small town in Tahiti or Raiatea, unexpectedly emerging from the past in front of us.

Two unrelated portraits struck me with their similarity of focus and composition, even if the two different styles had not made this similarity immediately obvious. The first painting is John Webber's *Portrait of Poedua, daughter of Orio* (1777), now in the National Maritime Museum in Greenwich. The other is Paul Gauguin's *Two Tahitian Women* (1899), now in the Pushkin Museum of Fine Arts in Moscow. The same women also appear in the large frieze *Faa Iehe* (Tahitian Pastoral, 1898) in the Tate Museum, London, and again in *Rupe Rupe* (1899).

At first glance, Webber and Gauguin's paintings have little in common, apart from representing beautiful, young, exotic-looking women. Poedua stands against a vaguely tropical background and cloudy skies. Her dark wavy hair reaches her shoulders and she has flowers near both ears. She wears a pareu reaching almost to her breasts. Her left arm is folded across her waist, and the other one lies along her side, holding a flywhisk.

Webber and Gauguin's Portraits of Women

Gauguin's painting represent two young women, also wearing a pareu: one reaching up to the breast of the central figure, the other draped over one shoulder of the other woman, leaving one breast exposed. Let us leave aside this second figure and concentrate instead on the central one, shown with her arms crossed and carrying a flat basket filled with mango petals.

Neither picture offers a salacious representation of Tahitian womanhood. In one, Poedua is obviously posing for a formal portrait; in the other, the two women seem to be quietly waiting for something. Certainly the styles and the palettes are different and do not suggest any particular resemblance. Yet, they are similar in one way. Each woman is more or less situated in the middle of the painting, with either light, colour, or prop used to emphasize a particular aspect of her anatomy. The composition of the paintings and the stance of the subjects actually turn each woman's breasts and belly into the main focus of attention.

The same composition occurs in both pictures, with horizontal shoulders, vertical upper arms, and folded horizontal forearm(s) actually creating a torso frame within the larger frame of the painting. While the women's hips and bellies are highlighted through a contrasting colour, it is really the breasts that occupy central stage through this particular inner frame.

Gauguin even goes further in this emphasis by suggesting that the breasts are being offered, served on top of the flat basket filled with flowers. *Two Tahitian Women* is the painting's official title, but it is also called *Les Seins aux Fleurs Rouges* (Breasts with Red Flowers), confirming this interpretation.

In this perfectly geometric configuration, a square frame within a rectangular painting, sit two perfect globes, the focus of the sailors' delight, those ideally shaped breasts of the *vahine*. The women, who show little facial expression, are not merely objectified; they are in fact reduced to being symbolized by their breasts.

Poedua would have thought nothing of it and not understood the particular interest taken in her breasts. A Polynesian woman of high rank, she posed suitably dressed for a portrait, her breasts revealed and her body otherwise clad in a draped pareu. It was common for women to bare their breasts in sign of respect, and there was nothing inappropriate in either her stance or her costume.

However, between her and Gauguin's models, the missionaries had established severe rules of dress and comportment to ensure women's modesty, wrapping them in "Mother Hubbard" gowns, and Gauguin's painting were clearly defiant of those rules. He was well known for flaunting all conventions in his dealings with native women and being an avowed opponent of the missionaries' influence. While Webber painted under a veil of classical references the figures and activities of people innocently close to nature, Gauguin's own depiction focused on the unbound vitality and vivid colours that also reflected his epoch's taste in the exotic. Two eras, two influences, two audiences, but the same focus.

We cannot leave the topic of painting without mentioning Bobby Holcomb (1947-1991), musician and painter, Tahitian by adoption, and French Polynesia's "Personality of the Year" in 1988. Holcomb was only interested in representing Tahitian life, customs, and dreams, so the women in his paintings were involved in ordinary or extraordinary activities to the same extent and in the same manner as men were, without explicit sexual connotations. Several of his paintings are included in a recently published anthology of writing and art, *Varua Tupu*.

The "Wifies"

Taatamara, Rarahu, Tuimata were some of their names, the Tahitian women who caught their white men's hearts. At least those who were celebrated in their lovers' writings, for who knows how many, over the centuries, had made it hard for these men to leave Tahiti and return home.

Taatamara was the woman Rupert Brooke fell in love with and called Mamua in his poem "Tiare Tahiti," the woman who lived in "this side of Paradise." It is believed that he had a child by her. This is worth mentioning since the topic often arises in Tahitian literature. It is also reflected in Tahitian society and population, and confirms the familiar White Man/Brown Woman pattern of relations: arrival, seduction, and departure, often leaving a child behind.

While the faces of the *vahines* are many and their personalities are variously alluring, their characters and behaviours usual fall within the same pattern. Let us consider here two books in particular, *The Marriage of Pierre Loti,* first titled *Rarahu, A Polynesian Idyll* and a less known work, Bjarne Kroepelien's *Tuimata*. These early books represent a literary genre of exoticism and escapism that is the very antithesis of modern Tahitian writings, more prosaically focused on the strong bond with the land, family life and connections, the reality of economic survival, and everyday life.

Rarahu

The Marriage of Loti is, in the words of Kaori O'Connor, who prefaced the 1986 edition, "a tale of amorous dalliance set in Tahiti, the exotic romance par excellence." A century after its first publication, it still speaks to the same desire for exotic settings and passionate affairs that European literature exalted at the end of the nineteenth century.

The romantic plot of *Le Mariage de Loti* is simple. An English naval officer by the name of Harry Grant falls in love with a Tahitian woman, Rarahu, his "little savage friend," during a stay in Tahiti. After their affair is over, Rarahu's ultimate fate is dramatic. She returns to Bora Bora, where she will die of alcoholism and consumption, barely eighteen years old.

In reality, the love affair is loosely based on several Julien Viaud had with various Tahitian women during a two-month stay in Tahiti. Viaud, a French naval officer, had been given by these women the Tahitian name

Loti (perhaps pronounced Roti, meaning rose, but not a particularly attractive sound to a French ear) and thereafter adopted the name as a literary pseudonym. How much of the book was true? Loti wrote in his *Journal Intime*, "I have only respected the truth in the details; the story itself is not true; I combined several real characters to compose a unique one: Rarahu, who seems to me a rather faithful representation of the young Ma'ohi woman." Our modern sensibilities are repelled by his description of such affairs. "Beautiful creatures, those Tahitian women—not classically Greek as to features, but with a beauty of their own which is even more attractive, and antique figures and limbs! Mentally, incomplete creatures whom one loves like fine fruit or fresh waters and gorgeous flowers" (1991:214).

The book so enchanted the imagination that there was an attempt at transposing it in 1898 into a lyrical work, titled *L'Ile du Rêve, idylle polynésienne en trois actes*. In it, Harry Grant, the hero of the *Marriage*, becomes Georges de Kerven and is now a French naval officer, and Rarahu is now named Mahenu, while the hero is still called Loti. This story had previously inspired Léo Delisbe for his opera *Lakmé* (1883), in which a British naval officer falls in love with an Indian woman, with disastrous consequences for her.

It would not be the only time Loti's writings would use the theme of a European naval officer falling for a native woman, going through a form of wedding, and eventually abandoning her and leaving her to die of love. His first book, *Aziyadé*, relates a similar story, loosely based on his affair with Hakidjé, a woman he met in the Levant in 1876. Then, in 1885, while briefly stationed in Nagasaki, he met and fell in love with a young Japanese woman, Okané-San, who became the heroine of his book, *Madame Chrysanthème* (1887). The story inspired another opera, Puccini's *Madama Butterfly* in which, this time, it is an American naval officer who loves and betrays a Japanese woman.

It is obvious why Rarahu, the first of this kind of heroine, is often seen as a prototype and has been described as one of the many romanticized avatars of the native woman. Dashing naval officers were, of course, the only young Europeans or Americans who could realistically meet women from so many exotic countries and legitimately leave them behind when duty called and their ships sailed away. In the words of Lesley Bland, Loti's biographer, the ever-occurring pattern was "landing, loving, leaving."

Although the native women in Loti's books were all beautiful and faithful, they were usually abandoned and often left to die because their white lover had a white fiancée at home whose own virtue (and above all, whiteness) could not possibly lose out to the brown native "wife." It was yet another area where one could not argue with the white superiority. The period, 1880-1900, was one when European imperialism was at its peak, and when it was admissible for Rudyard Kipling to describe the African as "half-savage, half-child."

The literary romantic exoticism favoured by some writers, particularly Loti, resonated with the readers. Thanks to the earlier eighteenth-century works, the French public was quite eager to recognize in *The Marriage of Loti* what they had already read, and there was nothing unexpected for them in love stories depicting exotic women in magnificent natural surroundings, beautiful and loving women who could be left behind with apparent impunity.

Loti must certainly be seen as another important contributor in his own century to the perpetuation of the Tahitian myth. He made Tahiti once more fashionable. He reports in his *Journal* that Juliette Adam (the director of the revue in which the *Marriage* was first serialized) told him, "In Paris, many people only write to each other in the Tahitian style, and end their letters with, 'I have finished my little speech. I salute you.' Apparently, there are also *Loti* candies, *Rarahu* ribbons, etc." (Vercier, 1991:14). With each book and each adventure, his reputation grew, but the cornerstone of his work was *The Marriage of Loti*. By 1905, it had reached its seventy-fifth edition, and in July of the same year, in a conference titled "Loti's Legend and the Truth about Tahiti," the speaker declared, "One no longer goes to Tahiti without having first read the idyll of *The Marriage of Loti*; and civil servants, officers, colonists, soldiers, and sailors all want to experience there the legend of the young ensign [Viaud-Loti's grade in the French navy when he was posted in Tahiti]" (Vercier, 1991:16).

Finally, if we needed to be even more convinced of his influence, in 1932 a specialist of the culture of Oceania, André Ropiteau, created an association to build a statue of Loti in Tahiti. "Seeing that Tahiti and all the islands of French Oceania have a moral duty of gratitude towards the great artist who so well described their charm, who made them so well known and loved so much, Pierre Loti; seeing that the best manner to express this gratitude would be to erect in Papeete a very simple monument: a bust of the young Navy officer set on a stele decorated with a head of Rarahu, the whole thing surrounded by flowers, in as quiet and Tahitian a corner as possible, such as the crossroads of Fautaua" (Vercier, 1991:16). Indeed, the bust was installed there in 1934, up the Fautaua Valley, where it still stands near the Mormon Temple and not too far from the Kanti Chinese Temple. The lovely natural site Loti had enjoyed no longer exists, replaced by the degrading signs of civilization.

It is probably true that for most of the writers who have celebrated a particular Tahitian woman, the island itself is equally the focus of their admiration and their love. Although Loti is sometimes disappointed and bitter in this book, he is passionate about the natural beauty of *"l'île délicieuse."* He expresses sentiments that will soon become fashionable among later writers about Tahiti. "Civilization was too present, our stupid colonial civilization, all our conventions, all our habits, all our vices, and the savage poetry went away, with the customs and the traditions of the past" (1991:52).

Tahitian writer Chantal Spitz hates the image of Polynesian women embodied by poor Rarahu, her "other self," even if the cliché celebrates their beauty and exotic charm, because she too is tainted by the same mythic depiction. "She sticks to my skin like a label," Spitz writes, a label she cannot remove. "I do not want to be a myth… [but how can I not be one] when all this time I have been described told spoken imagined sung fantasized." Rarahu is "tattooed onto my soul my identity my humanity my difference" *("Rarahu iti e autre moi-même")*. Spitz, a modern writer engaged in creating a new image for Tahitian people through her own literary work, cannot forgive Loti for having written a banal love story between a sailor and a "common girl" that has been stamped in the European mind and continues to exist there and move people's imagination.

A subplot in *The Marriage* is an equally relevant theme in Tahitian life and literature, and reappears with the second book we will consider, *Tuimata*. It is that of the children of a Tahitian mother and a European father, the *demis*, who today figure importantly in both Tahitian society and literature. It corresponds here to a real search by Loti (or rather, Julien Viaud). His older brother, Gustave, had died in 1865 after spending nearly four years in Tahiti and living with a Tahitian woman. The search concerns two sons that Gustave would have had, one in Tahiti, the other in Moorea.

Tuimata

The plot of the next book, *Tuimata,* is not original. A European man and a Tahitian woman love each other, he leaves, and she dies. Bjarne Kroepelien was a young Norwegian born in 1890 in a family of wine merchants, who went on a sort of Grand Tour and visited the United Sates, then Tahiti. He happened to be in Tahiti during the terrible 1918-19 Spanish influenza epidemic, from which died the young woman he loved, called Tuimata in his eponymous book. She was actually Amélie Teraiefa Teriierooiterai, the daughter of the chief of Papenoo and mother of his son, Bjarne Kroepelien Teriierooiterai. After returning to Norway, Kroepelien supported the child financially, but he never went back to Tahiti.

Kroepelien resumed his business life in Norway, but also collected everything that had been published about Tahiti, establishing a firm reputation as a collector and cognoscente of all things Polynesian. He also wrote several articles on the islands and in 1939 produced the French translation of some of Sparrman's reports during Cook's second voyage. Kroepelien then devoted his time to bibliographical studies and remains recognized as an expert in this field.

Tuimata is an exception in his literary output. During World War II, a number of years after the events the book relates, he wrote the story of their love affair. It was first published in 1944 and reissued several times. In the difficult days of the war, the public obviously enjoyed tales of dream, evasion, and exoticism, which did not please the Nazi occupants, who had the book withdrawn. In 1946, a Danish version was published and, through the joint decision of Johan Frederik Kroepelien, his Norwegian nephew, and Jean-Claude Teriierooiteraii, his Tahitian great-great-grandchild, the book was translated into French and published in Tahiti in 2009.

The story itself is conventional, even if more or less true. The young Scandinavian saw a beautiful Tahitian woman—in fact, he spotted her in the very place where Loti had first seen Rarahu, and the same enchantment ensued. However, there is more to the book than a love affair. Kroepelien is aware of the local history and culture and is able to provide vivid descriptions of the people he meets, completely unadorned of the romanticism that sometimes afflicts Loti's prose. His small portraits amount to a collection of characters; they are concise, sharply observed, and never harshly critical.

Teuru

Among various South Seas islands beauties, we meet James Michener's Teuru in *Return to Paradise*. She is a departure from the Tahitian women we have seen so far, and no longer the submissive love interest of a white man in search of exoticism. Beautiful young Teuru from Raiatea is to go to Papeete because "the Americans are back!" It is a tradition issued from the war, and her family's plan is for her to connect with a rich American. She will then live on his yacht, eat her meals at the yacht club, Chinese girls will sew her dresses, and she will wear shoes. Then, she can find herself a "handsome young lover," put him up at the Hotel Montparnasse, where she can meet him at night, "when the rich American is asleep."

During the war, explains her father, "the lucky girls who did reach the American camp [in Bora Bora], they lived in Paradise." A true Paradise, indeed: "Canned food. Jeep rides. Trips in the airplane. Whole cartons of cigarettes." Promises were also made by these Americans. "Come peace, Baby, you'll see me back here with a fist full of dough!" (1951:79-83). Young Teuru hesitates, but complies under pressure. However, once in Papeete, instead of the rich American she was instructed to meet, she faces three very different white men.

The first is an idealist Frenchman, Victor, a young poet who had always been drawn to Raiatea, and who stands enchanted upon his first sight of the island when both he and Teuru return to her home. He proposes marriage and cannot understand Tereu's shock at this breach of tradition. Disappointed by her refusal, he returns to France (1951:91-101). His greatest disappointment, however, stems from the reality he faces once on the island. While he seeks traces of the "godlike subduers of the Pacific," he only sees Teuru's degenerate father. Nowhere can he find "a trace of the grandeur that had once inhabited this land." No longer are the people beautiful, "their teeth falling with the white man's food, their health ruined with white man's diseases." He falls with what Michener calls "today's illness," which occurs when facing what has become of a beloved past. Victor is sickened to see "the sunny islands of Polynesia now owned by Chinese and Frenchmen and Germans. Everyone prospered. Only the Polynesians withered."

This is not an uncommon sentiment among the white men who arrived on the island in the pursuit of their dreams. Pierre Loti saw the island's "savage poetry" destroyed by a "stupid colonial civilization." Gauguin shared the same nostalgia for past he never knew. "A profound sadness took possession of me. The dream which had brought me to Tahiti was brutally disappointed by the actuality. It was the Tahiti of former times which I loved. That of the present filled me with horror" (1994: 15).

What the fictional Victor, Pierre Loti, and Paul Gauguin express is their disillusionment with the collapse of the Tahitian myth. They may have found love with the women—one side of the myth—but the harmonious life with nature they also sought in Tahiti—the second side of the myth—has vanished and they feel sadly betrayed. All three are French and it is ironic that the illusion of Eden, now ruined, was first perpetrated in Europe by a Frenchman, Bougainville.

Tereu's second white man is the English sculptor Earl Weebles. He is in poor health and, despite her care, he soon dies (Melville, 1951:103-109).

After each lover, she goes back to Tahiti. On her last visit, she finds her third white man, Johnny Roe, a derelict alcoholic American she brings back to health by once more returning with him to her island. She gets pregnant and, delighted, arranges to have her child adopted by a local Chinese merchant. The young American turns out to be the scion of a wealthy family, and all could go well for Tereu and her child, but the clash of cultures is such that he too, like her French and English lovers, will leave her behind as he goes away (1951:111-119).

In a sense, all three white men have followed the tradition. They have come, loved, and left her, and one made her pregnant, but there stops the similarity. In all three cases, there seems to be a genuine love story between each successive man and lovely Tereu, but we can see that she is different from the Tahitian lovers dreamed up by Loti and Kroepelien. She is no less passive, and she is so profoundly anchored in her culture that she does not understand what those three different men want from her. What she represents is more than herself, a young, beautiful, and desirable woman, but herself as the embodiment of the white man's island dream, as revealed in the dialogue between Teheru and the "honest-to-God American" Johnny Roe, who had always wanted to see the islands. She asks, "Why is it that so many white men want to come to Tahiti?"

"You've got to have somewhere you want to go."

"What did you want to find?"

"You, I guess."

James Michener knew the rebellious history of Raiatea, its inherited hatred of the French, and the traditions and customs of the islands. Thus, when the young Frenchman (despised by her father and many other islanders) asks her to marry him, Teuru is shocked. "After all, she was only seventeen, not nearly of an age to marry. She had not yet lived with a man, she had borne no children, knew nothing of life." Later, she explains, "All girls give away their first babies. How else could they get married?" Indeed, babies are necessary to demonstrate a woman's fecundity. She can then marry, having given her first child, as custom required, to be adopted by relatives or neighbours, where it will be lovingly cared for.

It is only within her family and friends, all far more closely connected than appeared at first sight (through the same system of adoptions), that she finds peace and understanding. She will eventually decide to marry the rich Chinaman who has already won the right to adopt her baby. All's well that ends well, and both Victor and Johnny Roe are left perhaps regretting their island girlfriend, but still shaking their heads at her incomprehensible behaviour. Teuru is a refreshing character, avoiding the forlorn fate of her literary sisters, and also happily escaping from the expected "wifie" condition.

The Wifie

What of these unions between white men and native women, a common enough occurrence when the balance of power is overwhelmingly in favour of white men? Indeed, it is a rich field for romance and exoticism, one in which the pattern of landing, loving, and leaving was more or less the norm. But what was the basis on which those unions were formed?

In "The Beach of Falesá," R.L. Stevenson described the "marriage" ceremony and the document drawn to attest that "Uma, daughter of Faavao of Falesá island of ----, is illegally married to Mr. John Wiltshire for one night, and Mr. John Wiltshire is at liberty to send her to hell next morning." The young "bride" wore "flowers behind her ears and in her hair she had the scarlet flowers of the hibiscus. She showed the best bearing for a bride conceivable, serious and still."

John Wiltshire soon felt ashamed of the deception. "That was a nice paper to put in a girl's hand and see her hide away like gold. A man might easily feel cheap for less. But it was the practice in these parts, and (as I told myself) not the least the fault of us White Men but for the missionaries. If they let the natives be, I had never needed this deception, but taken all the wives I wished, and left them when I pleased, with a clear conscience." Uma was indeed profoundly deceived. "She touched her bosom. 'Me – Your wifie,' she said" (1996:171-73).

With Gauguin, we have a different version of the "wifie," perhaps because the real Gauguin was less scrupulous than the fictional Wiltshire. Gauguin's Titi may have been less innocent than Stevenson's Uma, yet she was equally deceived. Gauguin described her ready to accompany him when he left Papeete, wearing "her very best dress for the journey. The *tiare* was behind her ear, her hat of reeds was decorated above with ribbons, straw flowers, and a garniture of orange colored shells, and her long black hair fell loose over the shoulders." We see the portrait he would have painted. He also showed how she betrayed a form of childish innocence, and how deceived she was. "She was proud to be in a carriage, proud to be so elegant, proud to be the *vahina* [sic] of a man whom she believed important and rich. She was really handsome, and there was nothing ridiculous in her pride, for the majestic mien is becoming to this race." He then compared the young Tahitian girl with a French prostitute, showing Titi to be the more honest participant in the games of love for purchase. It is as if her being used by a much older European is more forgivable because she is honest and because her race is prone to passion! "I knew very well that her calculating love in the eyes of Parisians would not have had much more weight than the venial complaisance of a harlot. But the amorous passion of a Maori courtesan is something quite different from the passivity of a Parisian cocotte—something very different! These eyes and this mouth cannot lie. Whether calculating of not, it is always love that speaks from them" (1994:146).

This rather glib description of the White Man/Brown Woman relationship is belied by the spectacle of the abandoned women watching their lovers sail away. He writes about another woman, Tehura, who "had wept through many nights. Now, she sat worn-out and sad, but calm, on a stone with her legs hanging down and her strong, lithe feet touching the soiled water." Tehura is not alone, for "here and there were others like her, tired, silent, gloomy, watching without a thought the thick smoke of the ship which was bearing all of us—lovers for a day—far away, forever" (1994:15). As we know, this image will be repeated often in the literature of colonial and exotic abandonment.

I deliberately merge fiction and reality when considering these native companions, the *vahines* of illegitimate unions, because the men who wrote about them were also deluded into believing a myth they felt to have been based on an earlier reality.

The Vahine and Tourism

The image of the bare-breasted *vahines* from the early days of contact is one that is still profitably cultivated by the tourist industry. Sixty years ago, Danielsson was already writing, "The South Seas are associated with love. Most commercial artists are well aware of this, and a hula girl waggling her hips always figures conspicuously on all travel agencies posters" (1956b:10). What worked then works even better now that tourism has grown to the scale we know, and this promotion reaches everywhere. Today, every tourist in Tahiti is handed upon arrival a map of the island and of Papeete figuring, among other advertisements, one from the Post Office showing a graceful dancer in full Tahitian garb, suggestively juggling with stamps and postcards issued by the Office. They are no more necessary to the sale of stamps than are our female models used to sell cars in similar promotions.

So lucrative and well received are the pictures of beautiful young women with golden complexion, almond eyes, and luscious black tresses that the models need no longer be Tahitian. John Stember, a photographer who markets pictures of native beauties, stated in a documentary *(Tahiti. La Vahiné)* that all the "girls" need is to have the right skin and hair colour. Once they meet these very basic conditions, they will do, provided they crown their heads with garlands, do not mind exposing perfect breasts, and are willing to pose in the surf in very scant attire. However, he found that there was "a lot of resistance here about nudity. It's probably their religious background." So, "trying to revive the old myth," he used models from Italy or Brazil, who have fewer scruples and much less modesty than the islanders, and who do not mind in the least "showing off their tits." They could easily be mistaken for local *vahines*, the mythical women of Cythera. These are the women I would call pseudo-*vahines*.

Miriam Kahn confirms that the women depicted as Tahitian on postcards or posters have sometimes little to do with Tahiti. In a 1995 interview, Teva Sylvain, another photographer of Tahitian cheesecake present in every postcard and magazine kiosk, showed Kahn a particularly seductive nude picture he had taken, and told her the model was French. "But I put the crown of leaves on her head, a garland of flowers around her neck, and a coconut-leaf basket in her hands to give her a Tahitian look. That's all it takes. Other than those props, there's nothing Tahitian about her" (2011:80).

And so is reborn the familiar figure of the seductive *vahine*. It matters little that she is a professional model with black hair and a great smile, as long as she seems to emerge both from the sea and from the depth of people's two centuries of imagination. In reality, photographers sometimes complain that real Tahitian women may not conform entirely to the canons of beauty desired to personify the Tahitian women of fiction. Moreover, they are often too shy to pose for provocative pictures.

Photographers' model or dancer, the *vahine* tourists see may have little to do with the average Tahitian woman connected to her roots and her island, yet she is the one on whom rests the burden of having to personify an ideal. A whole industry is based upon her, and she needs only represent a fiction. The message sent by travel agencies is that the Garden of Eden still exists and tourists can live as did the fictional islanders of their imagination in an exotic island made of dreams.

All these women, mostly concocted by Western imagination—the Cytheran beauties of the navigators, the illicit "wifies" of the next century, Gauguin's exotic models, the pareu-clad sirens of Hollywood, the pseudo-*vahines* of the tourists' postcards—all serve to incarnate and repeatedly reincarnate in their various guises a Arcadian myth still drawing us to the South Seas Islands.

Reversal of Desires

The sole objects of desire we have so far considered are the women abundantly described by charmed European males. In all these pictures of seduction, is there a place for *tane ma'ohi,* the Tahitian man? The former pattern of exchanges was limited to the conventional White Man/Brown Woman, now a trite configuration. However, several events have occurred that could have shifted this monopoly, including the arrival of women as part of the French workforce starting in the 1960s; the *service militaire* that sent Tahitian men to France to do their civic duty; and the always increasing number of tourists of both sexes. All these are opportunities for white women to meet *tane ma'ohi* and admire him, for he, too, participates of the same enchanted myth of Arcadia.

Among contemporary writers, Chantal Spitz is very vocal in rejecting Tahitian stereotypes elaborated by Europeans, particularly the *vahine* of the white man's dreams. To portray the first encounters between white men (those "strange brothers" with "their white skin and their different bodies," to whom Tahitians have just "submitted") and local women, she describes the golden-skinned women of her island, *vahine ma'ohi,* "made for love, made by love… dream of white men" by whom they were "always desired, sometimes loved" (2007:14-15).

So far, the evocation conforms to the navigators' experience and their descriptions of the women's tantalizing beauty. Where Spitz introduces a reversal is when the counterpart of that attraction is one these white men could not have anticipated. She writes, "Their women, pale, abandoned, with their different bodies, with their white skin, began in their turn to look upon our men." In her parallel poem, the white women now desire the Tahitian men.

This amorous inversion is evident in the comments made by a burly sailor on the *Aranui III*, a small ship that links the islands to Papeete. Up to two hundred passengers, mostly tourists, are taken monthly to the more isolated islands they would not otherwise be able to visit. One such is Fatuhiva, where landing takes place in front of the basalt peaks of *"la Baie des Vierges,"* (Bay of Virgins).

As an aside, the bay's name reflects two contrasting aspects of the arrival of the early Europeans. The *Baie des Vierges* may be a sanitized version of its other name, perhaps first given by eighteenth-century French sailors, *la Baie des Verges* (Bay of Penises, reflecting the shape of the peaks). The virgins' version is somewhat doubtfully related by Danielsson. "I thought at first that the name was meant ironically and had been given to the valley by some sorely-tried missionary after vain attempts to preach the seventh commandment to uncomprehending women, but the captain gave me another explanation. He pointed to the huge stone pillars which framed the bay and tried to show me how astonishingly like they were to statues of the Virgin Mary. I had difficulty seeing the resemblance myself" (1957:55-57).

The *Aranui III's* berthing in Fatuhiva is somewhat primitive and tourists have to be assisted in landing on wharfs or rafts, helped along by sailors, who grab them by the arms or the waist. One of them, a fine virile specimen, rather stocky in the Polynesian way, commented about the white women tourists. "They need that. To be held. They need to feel that strength, that smell of the skin, without perfume: a man's smell. It brings that to the tourists. We see that every year. Unbelievable. There is that, the charm of the islands, the charm of the people, of course, but there is also the charm of the sailors. It's more than exoticism, it's almost like an orgasm. That's the word!!" (Lefèvre, 2011). He then laughed, displaying magnificently hungry teeth.

The sailor is only saying what Spitz's poetry describes, with her particular praise of the sexual attractiveness of the Tahitian man, notably superior to that of the white man. "Tane Ma'ohi…Heaven's gift to love/Tane Ma'ohi/Desired by women/Loved in the darkest night/In the fare ni'au [thatched house]" and everywhere else on the island. This *Tane Ma'ohi*, the "envy of white men" is the "Dream of white women/Always desired/Sometimes loved" (2007:15-16).

This reversal, Brown Man/White Woman, suggested by Spitz, was unknown in the early days. The only European woman to visit Tahiti, Jeanne Barré, Commerson's female companion travelling with him disguised as his valet, is not known to have succumbed to the charms of the male islanders.[9] Spitz believes that tourism has now provided ample opportunity for white women to experience the appeal of Tahitian men. The latter connection may have emerged from real or anecdotal experience or it may only come from a general sense of deserved reciprocity, balance, and perhaps wishful thinking. What had once been sauce for the goose could now also be sauce for the gander. Very likely, the traditional requirement of faithfulness the white man requires of his woman is also at the base of *tane ma-ohi*'s boast of awaking the white woman's desire, a twentieth-century tit-for-tat for the events starting with Wallis's men two and a half centuries ago, and flaunted ever since in literature and art.

In 2012 and 2014, I watched Polynesian dancers and drummers perform for the tourists we were. While the women danced with a non-stop hip swirling motion (the arm movements seemingly less graceful and narrative than in Hawaii, but the pelvic energy apparently endless), the tattooed men's dance was extremely and almost aggressively masculine. When the dancers engaged in the traditional dance, their thighs open wide apart then smartly brought back together, each time more forcefully and with the occasional forward thrust of the pelvis, there was no mistaking the sexual nature of the dance. This was not lost on the white women in the audience, whose vocal response spoke of much appreciation. Traditional Polynesian dancing is usually appreciated by white male spectators because so much of it revolves around the female dancers' erotic motions, and there is little doubt that male dancers were equally up to the task of arousing the white female spectators' emotions.

There is no reason to assume that Tahitian men would not be attractive to white women, but to what extent the roles of sexual attraction have been reversed, or at least extended to include both sexes, is anyone's guess. In Chantal Spitz and Célestine Vaite's books, there is no doubt that this attraction exists. One of Spitz's male characters, Terii, and a Frenchwoman, Laura Lebrun, fall passionately and lyrically in love, a doomed love from which neither will recover. But this is Spitz's own style, full of passion and lyricism. Their love is doomed, not because of their ethnicity but because they represent two antithetic positions: the French nuclear expansion and the Tahitian opposition to it. Obviously, these cannot be reconciled and the two lovers must part. One unsettling detail in this meeting of the two races is that Terii has inherited his blond hair from his mother and her English father. As well, his French is so good that Laura, new to Tahiti, first assumes he cannot be native. Although she soon realizes her error, her original attraction remains ambiguous, since Terii had been at first glance mistaken for a European. Moreover, he is definitely not the run-of-the-mill Tahitian of the 1960s. After spending years studying in France, he has become an archaeologist.

One of Vaite's characters, Pito, now middle-aged in *Tiare in Bloom*, reminisces over his military service in France and the popularity of Tahitian conscripts with French girls, who "found them exotic, with their smooth chocolate skin." According to him, he only had to wink and the girls jumped on him.

However, *Tane Ma'ohi* has a long way to go before catching up with the *vahine* in becoming one of the leading Polynesian attractions. He certainly appears on the postcards and the tourist posters, but rather than merely posing or playing in the waves while offering his body to the viewer's admiration (as does the *vahine*) he is usually seen doing something both manly and "ethnic" (in his outrigger, fishing, playing his drum).

Would extending the amorous myth to include Tahitian men dilute and weaken the appeal, built specifically around an Arcadian myth with the *vahine* at its centre, a myth none of the other exotic and paradisiacal islands possess? Danielsson describes tourists going to Tahiti as different from the run-of-the-mill tourists elsewhere in the world. Rather than being interested in seeing monuments and engaging in the usual touristy events, they seek "the lost paradise," based on reports of the "natives' free and merry love life" in the South Seas. To illustrate his claim, he cites the titles of travel stories at the beginning of the twentieth century, such as *Tahiti, Terre du Plaisir (Tahiti, Land of Pleasure); Iles de Paradis (Paradise Islands); Les Iles où l'on meurt d'amour (The islands where one dies of love); Tahiti, Isle of Dream; The Island of Desire; The Island of Tranquil Delight; Voluptueux Sillages (Voluptuous Furrows)* and even more tellingly, *White Man, Brown Woman* (Danielsson, 1956b:9-11).

These earlier titillating titles given to novels are well matched by later titles of more serious books, such as Newton Rowe's *Voyage to the Amorous Islands* (1955), Danielsson's own *Love in the South Seas* (1956), or Anne Salmond's *Aphrodite's Island* (2010). I am not implying that the titles are not perfectly suited to the books' much more academic content, yet "amorous," "love," and "Aphrodite" refer to the same well established myth in the very same way as did the previous, lighter books. The hook is there, whatever the content, and the evocative titles are needed to reconnect with the myth that every reader would have recognized.

However, tourism has already changed the face of this traditional attraction to Tahiti by flaunting a new focus of love interest. Advertising itself (as do other similar tourists' havens, such as the Seychelles or the Maldives, as we will see later) as a honeymoon island, Tahiti has distanced itself from the *vahine*'s main function of making the white man dream—or, rather, it continues to do so by creating a propitiously erotic mood through her dances and her beauty. Having been replaced by a brand new bride, the *vahine* no longer needs to figure as the explicit object of the visitor's desire, but her image contributes the proper sexual ambiance of a *lune de miel*.

[IV]
CYTHERA REVISITED.

EVERY EIGHTEENTH-CENTURY VISITOR TO TAHITI shared a vision of what must have been, for sailors on those long harsh weeks or months at sea, the peak of their life experience.

However, a second look at island life revealed a number of problems. First, the contact between islanders and navigators had not always been smooth and had led to many skirmishes, the lasting scars of which are difficult to evaluate. As well, there were the darker areas of Tahitian culture that the French may have mostly missed and the English had tried to overlook, such as human sacrifice, infanticide, and cannibalism. Another source of puzzlement was also the relations between the sailors and island women. They seemed obvious and based mostly on trading sex for nails and other implements, but they were also based on the misunderstanding on the part of the sailors that all native women were promiscuous and available. Finally, there was the niggling problem of thievery, which the English resented but mostly tried to handle in a fair manner. So, was island life as idyllic as the new comers first perceived it to be?

Human Sacrifice, Cannibalism, and Infanticide

Human sacrifices and cannibalism were common knowledge in Tahiti, even if somewhat mysterious and often belonging to the domain of tales. They were also closely associated with warfare. Although known for the sweetness of their apparently hedonistic life, Tahitians at war were fierce and pitiless.

Salmond writes of Tahitian stories full of horrible details, of enemies "who were skinned alive... their skins worn as 'ponchos' in the field of battle; of women and children being killed in a horrible fashion; of captives being towed behind fleets of canoes until they drowned" (2003: 463). One chant says, "Woebegone is the man who would lose the battle! And who would be captured while fleeing! His head would be parted from his body, he would be abused, and his flesh would be torn away" (Akatokamava). Commesson, commenting on wars between neighbouring islands, wrote that they were "cruel, and no prisoners were taken. Both men and male children were put to death, while women and girls became the property of the victors" (Montessus, 1889:71).

Cook did not learn about human sacrifice until his third voyage; then, at his request, he attended one such ceremony. It was intended to placate Oro, the god of war, before an expedition to punish the inhabitants of Moorea who had revolted. Cook was accompanied to the *marae*, where the ceremony was to take place, by the surgeon William Anderson and John Webber, the artist who recorded the event, as well as Omai as a translator. This was the same Omai who had so successfully wooed English opinion during his visit to London, but no longer fared so well at home upon his return, having seemingly lost some of his Tahitian identity.

"A Human Sacrifice, in a Moarai, in Otaheite" (John Webber)

Cook understood that the victims were chosen among "low fellows or undesirables," either war prisoners or men taken from the ranks of the lowly *titi*. He was disgusted by the "bloody custom" and argued that such rituals would probably repel the gods rather than make them look propitiously upon the expedition.

Yet, Cook also accepted his limitations in attempting to make sense of what he saw; rather, he realized that he did not understand the Tahitian cultural context of what he saw. James Boswell reports, in a conversation with Cook, that the latter "candidly confessed that he and his companions who visited the south seas islands could not be certain of any information they got in the islands, or supposed they got, except as to objects falling under the observation of the senses; their knowledge of the language was so imperfect they required the aid of their senses, and anything they learned about religion, government or tradition might be quite erroneous" (Cameron, 1964: 78). Such regard for the truth and such humility in acknowledging one's limitations should be a model for all observers and would-be ethnographers.

As one would expect, there was a myth to explain the origin of human sacrifice in Tahiti. It is reported that a great drought had once occurred on the island, and the people asked the gods what curse was causing such a disaster. The king urged the priests to seek the gods' favour and beg for the rain to restore the fecundity of the land. They sacrificed pigs and offered fish and feathers. Still, the rains did not fall. A man was then sacrificed, and almost at once the clouds gathered and the rain started—a sure sign that the gods liked human flesh, and the god who seemed the most responsive to this type of sacrifice was Oro, God of War (Cameron:1964:113-15).

Many traditional cultures adopted cannibalism as a symbolic means of transferring *mana* or spiritual power from the vanquished enemy to the victorious warrior. The missionary William Ellis explained how he understood the connection between the sacrifice of cannibalism and the god for which it was intended. "Their mythology led them to suppose that the spirits of the dead are eaten by the gods or demons; and that the spiritual part of their sacrifices is eaten by the spirit of the idol before whom it is presented. Birds resorting to the temple were said to feed upon the bodies of the human sacrifices, and thus devoured the victims placed upon the altar... The king, who often personated the god, appeared to eat the human eye" (1859:358-59).

It is difficult to assess how widespread the custom was, although it was known to occur particularly in the Marquesas and among the Mahoris more frequently than in Tahiti or other Polynesian islands. Captain Furneaux, on Cook's second voyage, found evidence that some of his crew, sent to shore to gather wild greens, had been attacked by the natives in the Queen Charlotte Sound of New Zealand, and then cooked and eaten. James Burney, the ship's second lieutenant, sent to find out why the party had not returned, found "such a shocking scene of Carnage & Barbarity as can never be mentioned or thought of, but with horror" (Salmond, 2003:229).

In Polynesia, eating the body of man who was particularly despised or hated was also seen as the ultimate humiliation and revenge. Salmond explains that "to kill and ritually eat members of another group was the epitome of insult, 'biting the head,' an act which attacked their *mana*," and, quite specifically their capacity to perform effectively (2003: 2). Such deeds demanded retaliation, in a never-ending cycle of warfare.

The navigators saw signs of human sacrifice and cannibalism and were at some loss to reconcile what were to them abhorrent practices with the idyllic character they had assigned to the Noble Savage, partly epitomized by the Polynesian. Wisdom and experience prevailed and they had to acknowledge a practice they did not understand and otherwise rejected. Another later witness of cannibalistic practices was Herman Melville who, in *Typee* and *Omoo*, experienced firsthand life in Tahiti and the Marquesas. He too was able to accept the occurrence of practices alien to his own culture.

Equally disturbing to us was the presence of infanticide among the Tahitians, particularly when we appreciate the generosity and joy with which they greet today the arrival of babies among them. Infanticide was not a widespread practice. In fact, when the society was stable, it occurred exclusively among the *ariori*. However, there were times when overpopulation and scarcity of land and resources forced Tahitians to limit the number of births, and infanticide was the only known means of doing so. It is also important to know that for the Tahitians life only began when the baby drew its first breath. If the child breathed (for whatever reason, such as a mother's reluctance to kill it immediately upon birth), the newborn would then be safe to continue living.

The *arioi* were a religious group of travelling entertainers, whose leader, the chief priest of Raiatea, bore the title of *arioi-maro-ura* (comedian of the red girdle). They claimed to go back to the union of the god of fertility, Oro, and an earthly maiden, and were allowed to pass freely from group to group even when those were at war with one another. They celebrated free love and nature in their songs and dances, and commented on current events of local importance. They had many important functions in Polynesian society, which put them in a very special category with extraordinary privileges attached to their class. They constituted an exclusive group of warriors and orators, multi-talented artists known for their tattoos, their dancing, and their drumming. The missionary William Ellis admitted that "the *arioi* were a greatly respected society, combining the glamour of the secular stage with their religious functions." He also thought it was possible to see in infanticide an aspect of their religious dedication, perhaps similar to the commitment to celibacy in other faiths.

Their society was divided into several orders, each member rising from one to the next according to strict criteria. They ranged from the entertainers, mostly in the lower ranks, to those who held extremely privileged positions. Progress through the ranks was illustrated by the increasing complexity and numbers of their tattoos.

Novices were accepted among the *arioi* on the condition that any child they bore would be smothered at birth. George Forster, on Cook's second voyage, reports that the act was always done in secret. However, within the highest rank, newborns could be spared as descendants of the gods.

Among the justifications offered by the culture for such a practice was that the male and female *arioi* were recruited from among all classes and their promiscuity would have led to producing children from parents of unequal birth, a serious threat to the rigidly enforced lines of Tahitian social hierarchy. Moreover, their life was itinerant, as they travelled throughout the Society Islands, and bringing up children in those conditions would have been near impossible. Finally, the female performers who, like all *arioi,* were chosen for their attractive shape, could not jeopardize their performance and careers through childbirth and child care.

Low-ranking performers who may not have progressed beyond their lowly levels sometimes grew tired of their condition and resigned from the *arioi* society. They then rejoined their social class, married, had a family, and became respected members of the community. Danielsson believes that this reintegration into the more conventional model shows that the *arioi* culture was in fact contrary to the mores of the general population.

"The Allurements of Dissipation"

Upon his return to England, Captain Bligh was called to justify his actions relating to the failure of his mission and the mutiny on the *Bounty*. He did so by partly blaming his failure on the Tahitian women and easy living. "The women are handsome… and have sufficient delicacy to make them admired and beloved. The chiefs have taken such a liking to our people that they have rather encouraged their stay among them than otherwise, and even made promises of large possessions." Thus, it is understandable "that a set of sailors led by officers and void of connections… should be governed by such powerful inducement… to fix themselves in the midst of plenty in the finest island in the world where they may not labour, and where the allurements of dissipation are more than equal to anything that can be conceived" (1970:9).

The islanders and the navigators obviously had different criteria for sexual behaviour. Morenhout remarked in his *Voyage aux îles* that the Polynesians' innocent manners only caused scandal when Europeans saw them as disorderly.

Gauguin, who is not otherwise a compass for moral rectitude, would later share his insight into the natives' way of life, unsheltered from the elements, engaging in the same tasks; he believed that this commonality made men and women somewhat similar. Tahitian women bore little resemblance to European women who "thanks to [our] cinctures and corsets" have become artificial beings. Rather, Tahitian men and women shared many virile and feminine characteristics, which made their relations easier and gave their manners "a natural innocence, a perfect purity," as he put it. "Their continual state of nakedness has kept their minds free from the dangerous stresses which among civilized people is laid upon the 'happy accident' and the clandestine and sadistic colors of love… Man and woman are comrades, friends rather than lovers, dwelling together almost without cease, in pain and in pleasure, and even the very idea of vice is unknown to them" (1994:43-44).

The "continual state of nakedness" he mentions should be taken with a grain of salt, as there was usually an element of modesty in the islanders' dress. Women normally wore a skirt down to the knees, never exposing their genitals. Danielsson remarks, echoing some modern photographers, that women will not pose in the nude, and even "Gauguin, who used to offer liquor as a means of persuasion, often complained of the difficulty of finding models." Danielsson further comments that sex did not have quite the same meaning for Polynesians as it did for the Europeans, and modest women existed in the islands even before the arrival of the missionaries (1956b:61-62).

The Europeans saw the islanders' natural comportment and sexual behaviour as signs of extreme moral laxity. Even Cook, known to be sober and accurate in his reports (and also aware that he had not understood everything about the culture, rendering him for instance unable "to discuss Polynesian religious faiths") did not hesitate to provide his own perception of what was equally new and alien to him, the licentious mores he witnessed in the South Seas. "There is a scale in dissolute sensuality, which these people had ascended, wholly unknown to any other nation whose manners have been recorded from the beginning to the present hour and which no imagination could possibly conceive." Cook, although a superb observer, may not have been the best person to report on the matter. As James King, his second lieutenant on board the *Resolution*, put it, "Temperance in him was scarcely a virtue; so great was the indifference with which he submitted to every kind of self-denial" (Cameron, 1964:11).

Yet, he too held the island close to his heart. Upon leaving, he lamented his departure from "these happy isles on which Benevolent Nature, with a bountiful and lavishing hand hath bestowed every blessing a man can wish."

He was not only one to have ambivalent feelings. As many who had already commented on the lascivious dances, Arthur Bowes Smyth, surgeon on the *Lady Penrhyn,* remarked in 1788, "The greater part of the action during the Dance wd in England be thought the height of indecency, & indeed they seem calculated to excite venereal desires to a great degree, as we have reason also to think their songs are." While always censorious of the Tahitians' sexual mores (and with good cause, since he also witnessed the ravages of venereal diseases among them), Bowes Smyth could not refrain from saying upon leaving, "There cannot be a more affectionate people than the Otaheitans!" (Salmond, 2011:19).

There may be another interpretation for the enticing parade of lascivious women seemingly offering themselves to the white newcomers. This display was not always good-natured and sometimes followed incidents in which the two sides had clashed. Salmond describes a small skirmish between a few Tahitians accused of theft and some of the *Endeavour*'s crew attempting to intimidate them into returning the goods. The natives laughed the matter off and brought in a few women who then engaged in wanton gestures, "exposing themselves to the strangers," all activities avidly watched by the sailors. Salmond comments that, "although the British understood these performances as sexual enticement, in fact they were acts of derision."

The artist Sydney Parkinson also described contests between teams of women in Tahiti, with much twisting of their mouths, "straddling their legs, lifting up garments and displaying their genitals," the exercise being intended to humiliate the losers. When men became the recipients of such an exhibition, it was all the more demeaning for them because their sacred power had been defeated by that of the women. Salmond concludes in the case of the *Endeavour*'s crew, "Oblivious to these cosmological implications, the sailors responded in kind, exposing themselves to the women." However, they did not attempt landing and, as a further proof of the women's derision and demeaning intentions, they were pelted with fruit by the women, who also shouted at them (2009:51). While Parkinson, a Quaker, had been shocked at such antics, taking them at face value, James Morisson, an experienced sailor and boatswain on the *Bounty*, more realistically believed that those gestures had been made in the way of mockery. He felt that the girls who had exposed themselves were otherwise modest and timid, and did not allow any liberties from the Europeans.

There may have been other such incidents where local customs and rituals were completely misunderstood by the strangers. Beaglehole comments that if Cook's men were at times puzzled by the behaviour of the natives, the latter were often equally puzzled by the white men's own behaviour. While the nature of the sexual encounters was obvious, their performance was part of a tradition the Europeans did not know.

This is a common problem for all contacts between white men and any native culture newly encountered. While the Europeans' early dealings with indigenous peoples and the behaviours they observed were recorded as accurately as possible, we should be aware that their interpretations and analyses of the motives that drove them were usually of an ethnocentric nature and their knowledge of the language often superficial.

Mere gestures were often misunderstood in Tahiti, as they did not carry the same connotation as they did in England. A simple example will suffice to show the complexity of interpreting messages: When sailors saw people beckoning to them from the shore, they understood these gestures as an invitation to come closer, unaware that they were in fact an indication of rejection.

It is true that, according to most reports, the islanders' outlook on sexual matters differed from those of most European societies. In fact, the natives often succeeded in astonishing the British who witnessed some of their sexual games, particularly the explicit *arioi* celebrations that included genital displays and penis distortions on a spectacular scale. Monkhouse, Bligh's surgeon, witnessed such games and reported them in detail to his captain. Salmond relates Bligh's description of the *heiva,* the dance that had so shocked his surgeon. "The Heiva began by the Men jumping and throwing their Legs and Arms into violent and odd motions, which the Women kept time with, as they were conveniently cloathed [sic] for the Purpose, their persons were generally exposed to full view, frequently standing on one Leg and keeping the other up, giving themselves the lost most lascivious and wanton motions. Out of compliment the Women were directed to come nearer, and they accordingly advanced with their Cloaths up, and went through the same Wanton gestures which on their return ended the Heiva." Bligh then describes the extraordinary contortions and extensions of the dancers' penises, which left the British sailors amazed and probably somewhat nonplussed (2003:77).

There were other dances described by Monkhouse, although many have been lost nowadays after being banned by the missionaries. They had specific performers and specific functions. The *otea* was a man's dance, while the *hura* was danced by women of some rank. More ribald were the *timorodee* and the *upa upa*, the latter being danced by both sexes mimicking the sexual act. Such dances were interpreted by the European witnesses as having no other purpose than partaking in the purely wanton and lascivious nature of the Tahitians' mores. Indeed, such would have been the intent, had similar dances been performed in England by English men and women. We should be less convinced of the truth of this interpretation when in the context of Tahitian culture.

James Morrison, the same *Bounty* boatswain and later mutineer who had ample opportunity to get a closer look at the Tahitians' daily behaviour, warned that what Monkhouse saw at the *heiva* and reported to Bligh, could be interpreted differently. The gestures observed "are not merely the effects of Wantoness but Customs, and those who perform thus in Publick are Shy and Bashful in private, and seldom suffer any freedom to be taken by the Men on that account. The Single Young Men have also dances wherein they shew many indecent Gestures which would be reproachable among themselves at any other time but at the dance, it being deemed shameful for either Sex to expose themselves" (Salmond, 2003:77).

While the total sexual freedom reportedly in the South Seas islands may have been as much a Western fantasy as the reality, there is little doubt that Polynesian culture saw sexual intercourse as one of the major interests of life. Although sex was said to be practiced in private, it also seemed to dominate the thoughts, as there was a continual public reference to sexual activity. A preliminary to public meetings, for instance, often consisted of sexual jokes and references.

Commerson sensed that the picture of Tahitian society that would be relayed to Europe was likely to be censoriously received. As a man issued of two contemporary traditions, that of French *libertinage* and, more importantly, that of the belief in the innate rectitude of the Noble Savage's actions, Commerson considered how Tahitian sexual openness could be misinterpreted. "A prudish person would find nothing in all of this but a breakdown of public standards, a foul prostitution, the most shameless cynicism, but he grossly deludes himself in misunderstanding the condition of natural man… following without suspicion as without remorse the sweet impulse of an instinct always sure, because it has still not degenerated into reason" (Landsdown, 1964: 82).

The women's apparent amorality also struck Cook, who believed that Tahitians attached little value to chastity, particularly among the middle class, where a wife's punishment for adultery was a mere beating by her husband—a punishment probably not unlike that of a lower class woman's at her husband's hand in Europe at the time. He also saw that Polynesian men would offer young women, even their daughters, to strangers, but believed this to be simply a business proposition.

Commenting on the islanders' surprise at seeing their young women being rejected, as occasionally happened, Cook did not seem to appreciate that the European sailors' own behaviour had showed the natives that, for them, having sex with their women was the accepted way to proceed. This is what they would have expected anywhere in the world, even at home. John Mansfield, who studied maritime life in the days of Nelson, wrote, "It was not unusual for a monstrous regiment of women to march right across England so that they might join their mates on the other side." Prostitution in ports all over the world and in whatever era is neither new nor unusual. "In England, boatloads of women would come alongside an anchored ship of the line, and sailors would climb down the gangway, make their choice, and carried the women back on board" (Cameron, 1964:129-130).

The Tahitian women who pulled their canoes alongside the English ships with a similar intent were in no way different from these women. Francis Wilkinson observed, "The Women were far from being Coy. For when A Man Found a girl to his Mind, which he Might Easily Do Among so many, there was Not much Ceremony on Either Side, and I Believe Whoever Comes here hereafter will find Evident Proofs that there are not the first Discoverys" (Salmond, 2003:49).

Perhaps what made it appear different in Tahiti was the charm and grace of the women, the apparent naturalness of their behaviour, and the fact that it did not quite entirely look like prostitution. The island was naturally turned to matters of the flesh, and sex was not the evil it was often depicted elsewhere. Neither was it unimportant, and sexual education was guided by more experienced teachers who were expected to be very direct in their instruction.[10] It was simply a candid approach to sex, which the Europeans found surprising and made much of in their writings about Tahiti. Parkinson, the young Quaker naturalist, felt that "young girls are bred up to lewdness" and Danielsson confirmed that, "alongside beauty, sexual ability was the quality which the Polynesians valued most highly" (1956b:129). It would also seem that beauty and sex were intricately connected. In a recent French documentary (Lefèvre, 2011), a Marquesan comments that "when they landed on the islands, the missionaries were frightened by the women's nudity, and their sexuality, offered without taboo, was unbearable to them." He hesitates a little then adds, "Beauty is what? It's not the girl's appearance—it's ... before, a girl's beauty was the [hesitating, embarrassed] vaginal part of the woman. They were naked before. The most beautiful girl is not designated by the king or the queen but by the people. It's the population who will tell the king or the queen 'We don't want that one. Her buttocks are not symmetrical,' or something like that. Beauty is *"la chatte de la femme"* [a woman's sex]. There was a dance to determine that and it's the one religion [the missionaries'] had forbidden. That's where it started, because in the eyes of religion, it is abominable, but for us, the Marquesans, it's our pride."

It would have been impossible for newcomers, sailors eager to find sexual satisfaction after months at sea, to discern much difference among the women they met. They all seemed alluring and most of them were willing to entertain them, and the sailors could hardly guess at the social structure that distinguished between different types of women. More observant men soon realized that not all island women were promiscuous with the newcomers. William Wales, the astronomer on the *Resolution*, offered a more nuanced appreciation of the nature of most of the women who had sexual contact with the crews, particularly after they started being paid in nails or in trinkets.

> I have great reason to believe that much the great part of these admit of no such familiarities, or at least are very careful to whom they grant them. That there are Prostitutes here as well as in London is true, perhaps more in proportion, and such no doubt were those who came on board the ship to our People. These seem not less skillful in their profession than Ladies of the same stamp in England, nor does a person run less risk injuring his health and Constitution in their Embraces. On the whole I am firmly of opinion that a stranger who visits England might with equal justice draw the Characters of the Ladies there,

from those which he might meet with on board the Ships of Plymouth Sound, at Spithead, or in the Thames; on the Point at Portsmouth, or in the Purlieus of Wapping. (Salmond, 2003:42).

What had been a more or less innocent trade (sex for nails) and involved only a certain type of island woman, soon seemed to become far more venal. Cook had noticed that when the *Endeavour* had first visited New Zealand, very few of the women would sleep with his men. But the people seemed to have abandoned all discretion by the time of his second voyage, and their menfolk brought the women along to the ship. Cook, who by then was used to this behaviour, was nevertheless worried about the continued evidence of the transmission of venereal diseases. A moral man, he abominated his compatriots' conduct. "To our shame [as] civilized Christians, we debauch their Morals already too prone to vice and we interduce [sic] among them wants and perhaps diseases which they never before knew." What Cook did not yet know is that another culture, the Nootka, met on his last voyage in search of the Northwest Passage would also suffer from the same mementoes left by his sailors and the ravages of venereal diseases.

Paradise on earth, peopled by easy-going natives, where the women loved the sailors because of their white skin, and the whole island was devoted to the cult of Venus? The contemporary reports would on the whole seem to concur and, over time, this popular depiction has taken root to create the Tahitian myth, certainly now supported by touristic interests.

Skirmishes

A number of modern writers have challenged the usual interpretation of the easy-going sexual practices that greeted the European sailors, in particular Alexander Bolyanatz. While not the first one to do so (he refers, among others, to the historian C. Hartlan Grattan, a specialist of the histories of Australia and of Tahiti), he made it his task to seek other reasons than the accepted ones for this sexual laxity that had so pleased English and French sailors and inflamed their contemporaries' imagination.

In *Pacific Romanticism,* Bolyanatz refers us to the initial contact between Wallis and the Tahitians, where the latter arrived "within pistol shot of the ship," as Wallis himself put it. Wallis went on to describe the first encounter, where a large number of native canoes gathered around his own boats. He feared an attack and signaled his boats to come back. He described the ensuing confrontation. "At the same time, to intimidate the Indians, I fired a nine-pounder over their heads." Then, while the "Indians" attempted to cut off his boats, he succeeded in avoiding most of the canoes, save for one that got too close and started "throwing stones… which wounded some of the people." Upon this, the officer on board fired a musket loaded with buckshot at the man who threw the first stone, and wounded him in the shoulder. As soon as they perceived their companion to be wounded, the other people in these canoes leapt into the sea, while the other canoes paddled away, in great terror and confusion (Hanbury-Tenison, 2005:403-04).

Soon after this episode—full of terror and confusion for the natives—their women started trading sex for nails. Nine months later, upon Bougainville's arrival, the same "hospitality" was immediately shown his sailors. Bolyanatz argues that "the sexual aspect of the Tahitian response to Bougainville's landing in Tahiti was more a protective stratagem than a warm welcome" (2004:ix). He posits that Tahitians, or any other natives first meeting Europeans, acted as anyone would in similar circumstances. Their reaction to initial force and potential violence was, given the inequality between the two sides, to put forward all that would pacify the visitors. As well as sexual intercourse intended to assuage the white men, there was the extraordinary bounty of pigs, fowls, and fruit daily brought on board in amounts never seen before.

The encounter with Wallis was probably still smarting and the lesson had been well learned. Not surprisingly, they would then exhibit with the next set of visitors the same behaviour that seemed to have appeased Wallis's men. It could be seen as friendliness rather than as what it perhaps was in reality: pure self-protection. However, most Polynesian islands have always had a reputation for hospitality, which could take many forms, including sleeping with the hosts' daughters in more ingenuous days.

To Wallis, at least, it seemed that the Tahitians' friendliness was genuine, particularly that of Purea, the woman who bestowed her affections on him. When the *Dolphin* left, she accompanied the ship far out to sea and finally was forced to turn back "with such tenderness of affection and grief," wrote Wallis, "as filled both my heart and my eyes." Danielsson, who settled for a while in Polynesia in the 1950s and was eminently knowledgeable about island life and customs, has not, to my knowledge, suggested the scenario proposed by Bolyanatz.

It was neither the first time nor the last that an initial encounter between navigators and Pacific islanders had had unfortunate consequences. Salmond (2008:4-5) relates a few unpleasant experiences. The first concerns the Spanish Alvaro de Mendaña, discoverer of the Solomon Islands in 1567 and the Marquesas in 1595. As the crew rejoiced seeing land, they were surrounded by seventy small outrigger canoes carrying some four hundred naked, heavily tattooed men who then boarded the ship. When told to leave, they ignored the order. Mendaña had a cannon fired, upon which the terror-filled Marquesans jumped into the sea and swam back to their canoes. The only injury occurred when one of the islanders had his hand slashed with a sword while clinging to the side of the ship. The Spaniards calculated that, during the violence of the ensuing fight, some two hundred islanders had been shot. The expedition had already suffered several setbacks and ended in disaster. Yet, even during these violent encounters, some Marquesas befriended the Spaniards. Quiros, Mendaña's pilot, puzzled and unable to communicate, explained that "our men were very well received by the natives, but it was not understood why they gave us such a welcome, or what was their intention."

Another incident involved the Dutch explorers Willem Schouten and Jacob Le Maire in Tonga in 1616. Upon their arrival, the Dutch, whose ships carried little white flags as signs of peace (one cannot help but be charmed by such belief that white flags would be universally understood to denote peaceful intentions) found themselves confronted by nine or ten Tongan canoes and several larger ships. While the Dutch went ashore, their ship was attacked and in the ensuing scuffle, a Tongan chief was shot. In retaliation, the Dutch ship *Eendracht* was violently rammed at high speed, a common Polynesian tactic, and the encounter led to many natives being killed.

An instance of such violent experiences (Salmond, 2011: 21-41, 63-64) greeted Cook's arrival in Managaia on the *Resolution* and the *Discovery*, with Cook apparently behaving with unusual harshness. Another disastrous incident occurred between Captain Furneaux, on the *Adventure,* and some local Maoris, resulting in the massacre of some of the crew. Often, such events were the result of misunderstanding. For instance, the encounter in Hawaii that resulted in Cook's being killed in a skirmish was based on events and beliefs that made sense to the Hawaiians. Cook, coming from Tahiti into Kealakekua Bay, "the pathway of the gods," was seen as the incarnation of Lono, the legendary ancestor and god of fertility, who had sailed to the island of Kahiki (Tahiti) and was deemed to have returned in the form of Cook. Further events soon showed that Cook had not followed the path expected by such superstitions, and the English were no longer welcome ashore. As they returned once more to replenish their provisions and make repairs to their mast, a fight occurred during which Cook was killed (Salmond, 2003:409-14). He was only identified through his mangled hand, the result of an explosion at a younger age.

In 1787, during Lapérouse's attempted circumnavigation, one of his ships, the *Astrobale,* was attacked by Samoans, who killed some of the crew and the ship's captain. Lapérouse himself and his own ship, the *Boussole,* presumably disappeared near the archipelago of Vanikoro the following year. His had been a purely scientific and economic expedition, with on board the astronomer Dagelet, the geologist Lamanon, the botanist Lamartinière, three other naturalists, and three illustrators. Hardly the composition of a warship brashly intent on conquest.

Salmond rightly points out in her *Voyaging Worlds* that many such skirmishes, some more serious than others, that opposed Polynesians and Europeans, were between fighting sailors who misunderstood one another. They were separated by language and customs. Upon their arrival, the European ships were usually met by native canoes, often in large number and usually driven by mere curiosity. However, should even one of their men be harmed by the visitors, they would immediately resort to their usual response: hurling stones with slingshots and trying to ram the foreigners' ship with their canoes. These canoes, while ominous-looking to other Polynesians, had little power against the explorers' ships and their guns. With firepower and lethal range far superior to the natives' weapons, they would naturally gain the upper hand. In this context, Bligh's arrival, the one discussed by Bolyanatz, only followed a natural and tested pattern.

Thievery

"No sooner that you appeared among them than they became thieves," had warned Diderot. It would soon be established that the Europeans' tools, weapons, cloths, and ornaments would prove irresistible to the natives. Some of Cook's officers, while otherwise admiring the island and the people, noticed the native males' unfortunate tendency to appropriate small objects. "The island appeared to us to be the Paradice to those Seas, the Men being all fine, tall, well-made, with open countenances; the Women beautiful… the Middle size, zingy, supple figures, fine teeth and Eyes, and the finest formed Hands, fingers, and Arms that I ever saw… And though the Men were inclined to steal little things from us, yet the Women seemed free from this propensity," noted Lieutenants Elliott and Pickersgill, on Cook's second voyage (Salmond, 2011:203).

Generally speaking—and such "thieving" was found with many native groups—thefts would occur whether or not the native populations' cultures were based on sharing goods. A man, taking for himself, his family, or his community anything that rightfully belonged to well equipped newcomers could only be seen by the latter as a thief because this is how such action would be considered in their own culture. Indeed a whole penal industry already existed in Europe, where all crimes, from debts to murder, was codified and sanctioned according to strict regulations. The natives did not necessarily see this manner of acquiring goods as the infraction the Europeans deemed it to be. Commesson understood the problem differently. "Since there is no right of property in nature, it is purely a convention. The Tahitian, who does not really own anything and offers generously whatever others desire, has not recognized the exclusive right of ownership" (Montessus, 1889:61).

Once more, we should be weary of interpreting from an ethnocentric point of view the actions of groups whose culture is still alien to us, since we do not share the same values and usually ignore the cultural context of these actions.

The references to "stealing" and "thieving" by Polynesians abound in the records of the time, yet these offences seemed to have taken little away from the pleasure of their perpetrators' company. Admiral George Anson, in 1765, noticed that the natives of the Gilbert Islands "were friendly but of a thievish disposition." Similarly, Cook thought that, although the islanders were "unabashed thieves," their behaviour and appearance were so pleasing that their islands should be called the 'Friendly Islands' (Tonga)."

Stealing appeared to be an unavoidable consequence of encountering desirable goods. Often, these goods had been previously offered as gifts or as trades for foodstuff, water, breadfruit, or any other item the Europeans desired. The casual transition from gifts freely offered to stolen property may not have been as clear cut in the islanders' minds as in those of the Europeans, particularly if the latter continued to help themselves to the island's bounty—a view supported by Commerson.

The accusations of theft were almost universal among the white newcomers. But this too seems a little more complex, particularly when we think of the coveted iron implements and weapons. Soon, the sailors started stealing these to secure the women's favours, a practice that started with Wallis's men, but was not limited to them. Cook's crew was equally adept at appropriating nails for the same purpose, and a certain scale had even evolved, whereby a lady's charms were to be matched by the size of the nails. So serious and disturbing was the problem for a captain that Cook ordered one of the thieves under his command to be given two dozen lashes for "theft of spike nails" (double the usual sentence for theft and disobedience), and in a spirit of fairness, the same heavy punishment to be given to another for "theft from natives" (Salmond, 2003:433).

There is little doubt that the natives made very free with whatever could be carried off. When one of Cook's officers returned to his cabin after a brief tour of duty, he found that not only had the woman presumably waiting for him to come back vanished, but so had his bed linen. Others resorted to devious tricks, such as throwing back out to sea so they could be sold again the coconuts the sailors had already purchased. While this was rather naïve cheating and thieving, there were many other occasions when they created serious and sometimes lethal confrontations. From mild irritation to enraged desire for retaliation, the natives' thieving in Tahiti was not without consequence.

It was one of the many problems the arrival of the white men had created, and honest men such as Cook felt the damage done to the natives by Europeans and "civilized Christians." "We debauch their Morals already too prone to vice and we introduce among them wants and perhaps which they never had before knew, and to which serves only to disturb that happy tranquility they and their Fore fathers had enjoy'd." To anyone denying this truth, Cook challenges, "Let him tell me what the Natives of the whole extent of America have gained by the commerce they had had with the Europeans" (Salmond, 2003:26).

However, there is a twist in what appeared to be the Tahitians' desire to acquire what belonged to others—after all a commonplace expedient in the history of exploration and colonization. Many Tahitians had an unexpected reason for doing so: Hito, the esteemed god of thieves, had a strong following among the islanders. One explained to William Ellis, a missionary, "We thought when we were pagans that it was right to steal when we could do it without being found out. Hito, the god of thieves, used to help us."

Tahitian thieves were known to use incantations to induce the other gods to turn a blind eye on their activities. They also believed in the use of incantations by their victims to reveal the thieves' identity. In other words, the petty thieving of which the navigators were victims was not specifically related to them, even if their having far more enviable goods (particularly iron) made them all the more desirable victims. This was certainly the case with Cook's quadrant, indispensable for the observation of Venus. Seeing the importance he attached to it, the Tahitians may well have believed they had acquired some treasure. Cook succeeded in recovering the parts, but the theft must have rankled.

While it was culturally acceptable to be a thief, it was only so if Hito's protection worked. When it failed and the thief was caught, severe sanctions were sometimes inflicted upon him by his Tahitian victims. Such was not usually the case with the navigators, particularly with Cook, who was determined to avoid unnecessary killing, even when subjected to constant pilfering.

A common occurrence was the one involving Lieutenant Pickersgill and Johann Forster at Tahaa, across the lagoon from Raiatea. While staying with the local chief, one of their bags was stolen. They, in turn, confiscated several items of value and took hostages, saying that these would be returned and the hostages freed upon the return of their bag. When an angry crowd gathered, they fired above the natives' heads, and everyone fled in terror. The stolen bag was then returned to the English, who gave back to their owners what they had confiscated, and the hostages were freed. In the face of those endless irritations, many among Cook's entourage were irked by what they saw as his own weakness in dealing with this constant thieving (Salmond, 2003:259).

The naturalists and scientists were equally disturbed by this behaviour. Solander and Monkhouse once had their pockets picked, losing an opera glass and a snuffbox. Banks so forcefully joined in their recriminations that the chief immediately offered goods of his own to make up for the thefts. When those were rejected, he hastily went after the thieves and recovered the stolen objects, momentarily restoring the peace. Banks admired "a policy at least equal to any we had seen in civilized countries, exercised by people who have never had any advantage but mere natural instinct uninstructed by the example of any civilized country" (Salmond, 2003:67). In spite of the disappointment they must have felt with these constant incidents, certainly not a trait of character they would have wished the noble savages to have, they still maintained some faith in the natural laws of primitive societies.

Others saw in the Tahitian's thieving habits a just return for the actions of the English. John Marra, on *Resolution,* remarked there were good reasons why the islanders took things from the English. "Is it not very natural, when a people see a company of strangers come among them, and without ceremony cut down their trees, gather their fruit and seize their animals, that such people should use as little ceremony with the strangers as the strangers do with them; if so, against whom is the criminality to be charged, the Christian or the savage?" (Salmond, 2009:266).

Indeed, for some, the islanders' habit of appropriating the officers and crews' personal possessions was deemed to be a fair response to the latter's behaviour.

Morrisson, sailing with Cook, saw it in this light. "It's no disgrace for a Man to be poor, and he is no less regarded on that account but to be Rich and Covetous is a disgrace to Human Nature & should a Man betray such a sign and not freely part with what He has, His neighbours would soon put Him on a level with the Poorest themselves, by laying his possessions waste and hardly leave him a house to live in—a Man of such description would be accounted a hateful Person" (Salmond, 2003:78).

This was written in 1792. The Tahitians would have fared even worse had they been "discovered" in the sixteenth century, when the notion of the noble and basically good savage had not yet been formulated in European minds.

Those are only some examples of the conflicts between Europeans and South Seas islanders. It is the unusual navigator's report that does not include a violent encounter with hostile natives. Gilbert (1973:222) describes the "familiar pattern of events: friendly approaches, misunderstandings, ambushes, kidnappings, and retaliations," as he reviews the actions of sixteenth-century Spanish explorers. When we compare them with the initial contacts in the eighteenth century, the main difference is that the later navigators were much better equipped and armed than their earlier counterparts.

Many have pondered over the events that occurred there, with two very different cultures facing each other in an unequal balance of power and technology. Our records only come from one side, relying on hard facts as well as the impressions that filled the white men's memory and imagination. We know what probably happened, since the arrival of white men on any native lands follows familiar patterns, but we do not really know how the natives experienced the Europeans' arrival, and how much fear and resentment may have been hidden behind the overt welcoming ways of the islanders. There are traces of undated epic poems, but the Tahitians left no written record to present their version of the events. However, some early missionaries had recorded the natives' impressions upon the *Dolphin's* arrival—the very first European ship and crew they encountered and saw as terrifying beings, armed with thunderous weapons that killed many with their smoking projectiles. It is no wonder that they sought to appease such frightening men.

Our ignorance extends to many other aspects of Tahitian life. Salmond (2009:12) deplores the loss of material gathered by contemporary scientists and missionaries. She mentions four in particular: Joseph Banks's account of Tahitian cosmology by Tupaia (1769), Máximo Rodríguez's account of Tahitian customs and vocabulary, written during his visit (1774-75), the Rev. John Muggridge Osmond's two manuscripts, one of Tahitian life, language, and customs, and the other of the history of Tahiti (1848).

All the reports that reached Europe offered a similar emotional vision of Tahitian life. Salmond notes that "the European explorers saw the islanders through a haze of their own enchantments"—a similar comment made about Gauguin's paintings and no doubt still valid for tourists who expect to see in Tahiti the constructs of their imagination. She reminds us that, even if the explorers who discovered Tahiti came from Europe "during the Age of Reason, fantasy was far from dead, and the worlds that came together in these meetings were as much imaginative as real" (2009:21).

What the reports related, with fewer of the provisos required for a more balanced appreciation (the odd serpent hidden in this Garden of Eden and already spotted by some explorers) was the natives' charm and their enchanting ways, set in a nature more beautiful and in a climate milder than most Europeans had ever seen. Little more was needed to establish the Tahitian myth. Earthly Paradise, New Cythera, the Elysean Fields, Thomas More's Utopia, the Golden Age, or Arcadia is how Otaheite was perceived and variously named, inhabited by amorous flower-women. The women were thought to be giving a gift, whether naturally offered through the customary Tahitian generosity and the apparent laxity of the mores, or sometimes completely misunderstood, or even induced by the potential danger presented by incomprehensible but well armed visitors. This gift of the women's sexual availability contributed to what is the essence of the typical myth: that it should be an unassailable cultural fact.

Yet, the Europeans, in spite of their evident liking for the islanders, brought to them all the foreseen evils, including diseases, missionaries and religious confusion, whalers and traders, slavery, depopulation, and the abandonment of customs and traditions. Tahiti was no more spared from them than were so many other parts of the colonized world.

PART TWO

RUINATION

"One day they [the white men] will come with the crucifix in one hand and the dagger in the other" (Denis Diderot, French philosopher and *Encyclopédiste*).

"Disease and gun powder is all the benefit [the Islanders] have ever received from us" (George Hamilton, surgeon on board the *Pandora*).

THE FILM *MUTINY ON THE Bounty* we considered earlier contributed enormously to confirming in the viewers' mind the reputation of Tahiti as an island of easy-going people and beautiful maidens attracted to the Europeans' white skin. While conforming to some aspects of historical reality, this film was most of all an exercise in wishful thinking. Of similar vintage, the movie *Hawaii* shows us a different aspect of the contact and post-contact periods. The situation depicted in two Polynesian islands, Tahiti and Hawaii, reflects the evolution we examine in this book. The

Tahiti of the *Bounty* is still the earthly paradise that Hawaii also was when the English ships first arrived. *Hawaii* shows the almost immediate consequences of these initial encounters. The two films constitute the before and after pictures of colonization in the Polynesian islands.

Indeed, were we to see *Mutiny* and *Hawaii* consecutively, as some people must have done at the movie theatre only four years apart (1962 and 1966), the devastation caused by the white man' arrival and presence in Polynesia might have made some of the audience boo Bligh's men as they sometimes do the appearance of the villain on the stage. Granted, the islands were different and a few decades had elapsed in the meantime—and the white man was no longer English but American—but one could certainly see in *Hawaii* the consequences already presaged in *Mutiny on the Bounty*.

The film *Hawaii* is based on the third chapter ("From the Farm of Bitterness") of James Michener's book by the same title. Once more, I chose to consider the film on the assumption that more people saw it than read the book. Since I am only concerned here with the images implanted in people's minds, the medium I choose is the one that I believe drew the greater numbers.

As with *Mutiny*, the quotes are taken directly from the film *Hawaii*, using three characters' declarations to encapsulate the devastating aspects of post-contact events: a converted Hawaiian prince addressing in 1819 a newly graduated seminary class to draw a history lesson; a whaling ship captain; and the doctor attending to the victims of a pestilence, "this dreadful plague… prepared to strike once more with demonic force, killing, laying waste, destroying an already doomed population. It was the worse disease of the Pacific: measles." With only minor changes, whatever pertains to have occurred in Hawaii in the film could just as easily have happened in Tahiti.

The Prince to the future missionaries:

> Only forty-three years ago, when the English captain Cook arrived in Hawaii, my brothers for the first time came face to face with a Christian world whose existence they'd never even dreamed of. We beheld your weapons, your tall ships, your way with books and numbers, and we could not doubt the greatness of your Christian gods. In less than one generation, we leveled our temples, burnt

our pagan idols, and stood like waiting children for the revelation you promised to send us. But instead of sending God's words, you sent adventurers to steal our lands, pestilence to ravage our countryside, strong drink to destroy the minds of our sons, and whalers to despoil our daughters and cast them aside like little wounded animals, to die in torment from that disease for which we had no name until you came.

We should note that Tahiti's conversion to Christianity was different from what we can gather from this film happened in Hawaii. But the history lesson could mostly apply to Tahiti as well.

The American whaling captain to a sailor who had assaulted a young woman, then to the missionary who had intervened in the attack:

"Rape is a rotten sport any time, but in Lahaina, it's a damn waste of energy. Now, get out of here and find yourself some cooperative ones." Then turning to the missionary: "My men are not sailors, they're whalers. And for over three years they've been on this voyage and they've haven't touched land in eight months. Now, the females of Lahaina have got the warmest blood in the Pacific, and my men have neither the strength nor the inclination to fight them off. And neither, might I add, have I."

The doctor, attending the people stricken with measles, to the missionary:

"When Captain Cook discovered these islanders fifty years ago, they were a true paradise. Infectious disease was unknown. They did not even catch cold. They were 400,000 and now they are less than 150,000. You and I may well live to see the last Hawaiian to his grave." Bitterly adding, "With proper Christian services, of course!"

Such is the topic of this Part II of the book, Ruination.

[V]
THE WARNINGS

THE FIRST CONTACTS WITH EUROPEANS had often been hostile. From the early time of the Spaniards sailing around the Pacific to the arrival of the British and the French in the eighteenth century, many islanders were killed in bloody encounters. The forces opposed were too extreme in the power they could bring to these skirmishes, guns against clubs and spears, and the contrast of cultures involved was too extensive not to cause extraordinary physical damage, social disruptions, and cultural upheavals.

Initially, the misunderstandings leading to violence were often of a cosmic nature. Those European ships invading Tahitian waters did not correspond in any way to the islanders' worldview, nor could the islanders make sense of them. At first, they were not even sure that the Europeans were human beings. As well, some local priests had actually predicted the arrival of a new kind of people who would change their lives forever, so it is not surprising that first contacts would sometimes reveal the disarray of the natives and their inability to grasp the nature of the events facing them.

James Douglas, 14th Earl of Morton, president of the Royal Society, had been shocked by the violence of the clashes with natives and had urged Cook and Banks, among others, "to check the petulance of the sailors and restrain the wanton use of Fire Arms." He reminded them that the natives were "the natural and in the strictest sense of the word, the legal possessors of the several regions they inhabit." Finally, he warned that, "should they in a hostile manner oppose a landing, and kill some men in the attempt, even this would hardly justify firing among them, 'til every other gentle method had been tried" (Beaglehole, 1974:150).

However, in the heat of the affray, often misreading the mood and intentions of the natives, surrounded by canoes, feeling the stings of the spears, and very likely fearing for their lives, the sailors found it only too easy to forget these injunctions. As well, the islanders' constant petty thieving and the sailors' reprisals caused hard feelings of resentment on both sides.

Cook, influenced by the Earl of Morton's directives (called *Hints*), had drawn a set of instructions to guide the exchanges between his crew and the islanders. He intended to do as little harm as possible to people inexperienced in dealing with white outsiders. His crew was "to endeavour by every fair means to cultivate a friendship with the Natives and to treat them with all imaginable humanity." Mindful of Wallis's experience and his crew's ill use of the ship's equipment, he also appointed someone to control all trades, particularly trades involving anything made of iron, with the specification that all items had to be traded exclusively for provision. Aware as well of the duplicity involved in the disappearance of material, he also warned the sailors that whatever they may lose on shore—iron nails, tools, or weapons—would be charged against their wages (Salmond, 2003:63-64). The sexual exploits of Wallis's sailors and the way they had been financed were already well known to the crew, and Cook's directives were not appreciated.

Cook's efforts to behave humanely and fairly are striking, particularly in his earlier voyages. Unfortunately, he and the handful of navigators first to reach the islands were only the prelude to numerous life-changing and culture-destroying arrivals.

There had been ample warnings that Europeans' meddling in the natives' affairs would be harmful and have dire consequences for people unaware of the woes that could so easily be unleashed upon them. One of the early and often quoted admonitions is the one issued by Diderot, one of the Enlightenment's bright figures, a philosopher whose prophetic warnings were soon to be realized. He enjoined Bougainville to steer his ship away from "these innocent and fortunate Tahitians," to whom he could only bring unhappiness. "You took possession of their country as if it did not belong to them… No sooner had you appeared among them than they became thieves; no sooner had you set foot on their land than it became stained with blood." When Bougainville finally sailed away, Diderot bemoaned the fate of "those good and simple islanders [who] bade you farewell. Oh that you and your compatriots and all the other inhabitants of Europe might be engulfed in the depths of the ocean rather than see them again." In the same *Supplement au Voyage de Bougainville* (1796), Diderot further urged the unlucky Otaheitians to cry, not at the departure of these "ambitious and evil men" as some did when the white men sailed away, but at their arrival. One day, he told them, the piece of wood tied to their belts [crucifix] and the iron in their hands will put you in chains, will cut your throats, and will subject you to their extravagance and their vices. Worse, he predicted, one day you will be like them.

Diderot further entreated Bougainville, through the voice of an old Tahitian, to sail away as soon as possible. "We are innocent, we are happy, you can only ruin our happiness… We are free, yet you are already burrowing in our land the title of our future slavery… This country is ours. This country is yours! And why? Because you set foot on it? If a Tahitian landed one day on your shores and carved on one of your stones or on the bark of one of your trees: This country belongs to the natives of Tahiti, what would you think?… Leave us to our way of life; it is wiser and more honest than yours; we do not want to trade what you call our innocence for your useless lights. All that is good and necessary, we already possess."

Bougainville himself commented on a wise old islander, perhaps the same one Diderot had imagined, who carefully stayed away from the Europeans and "whose worried expression seemed to announce his fear that those happy days he spent in leisure would now be disturbed by the arrival of a new race."

Lapérouse also anticipated the bleak future for the native populations faced with the white men's overpowering forces. He observed in his journal, "What right have Europeans to lands their inhabitants have worked with the sweat of their brows and which for centuries have been the burial place of their ancestors? The real task of explorers was to complete the survey of the Globe, not add to the possession of their rulers" (Moran, 2003:33). Married to a Creole from Île de France (Mauritius), Lapérouse was perhaps more sensitive to these issues than most people of his time.

George Hamilton, surgeon on the *Pandora* sent in 1788 to round up the *Bounty*'s mutineers, recognized that life in Tahiti had already been profoundly affected by the European visitors. "Happy would it have been for those people had they never been visited by the Europeans; for, to our shame be it spoken, disease and gun powder is all the benefit they have ever received from us, in return for their hospitality and kindness." His words were echoed by Williamson, third lieutenant on the *Resolution*, who, à propos of a somewhat puzzling incident with a stolen goat during which Cook seemed to have shown unusual severity towards the natives, referred to the latter as "these poor & before our arrival among them probably a happy people."

In England, the historian and politician Horace Walpole commented in 1780, "Not even that little speck could escape European restlessness!" Far from escaping the attention of those restless Europeans, that little speck of land, Otaheite, had been visited during the decade that preceded his comment by no fewer than eight ships from three nations. They, and the numerous ships that would follow, would bring all the elements of western civilization, including the expected practical changes (more efficient weapons, iron tools, better clothing) and the underground and deeply troubling consequences (venereal and other diseases, missionaries, corrupting transient whites, slavery). Even without being able to guess the details of what the ships would eventually bring with them, Walpole anticipated that European "restlessness" might have a devastating effect on these islands.

Most visitors, faced with the changes they could already discern, were aware of the impact their visits would have, even if they could not possibly foresee the extent of the future damage that would derive from them. There was even a touch of sanctimonious satisfaction for some at being more civilized than others in their treatment of the natives. Banks wrote that their connections with the Tahitians, "'who already known our strength and if they do not love us at least fear us" would permit them to "gain some knowledge of the customs of these savages; or possibly persuade one of them to come with us who may serve as an interpreter, and give us an opportunity thereafter of landing where we please without running the risk of being obliged to commit the cruelties which the Spaniards and most others who have been in these seas have brought themselves under the dreadful necessity of being guilty of, for guilty we must be" (Salmond, 2003:63).

However, it would not be the Spaniards, who had abandoned their territorial and commercial ambitions in the South Pacific, but the English themselves who would start the rapid process of destroying the Tahitian culture. Others would naturally follow.

Observers and writers continued to reflect on the effects of the contact between natives and European visitors. Herman Melville, witnessing the "wild grace and spirit of the Marquesan girls' dancing," was also confronted with the "grossest licentiousness and most shameful inebriety" of the sailors present at the dances. "Alas for the poor savages when exposed to the influence of these polluting examples!" he lamented. "Unsophisticated and confiding, they are easily led into every vice, and humanity weeps over the ruin thus remorselessly inflicted upon them by their European civilizers. Thrice happy are they who, inhabiting some yet undiscovered island in the midst of the ocean, have never been brought into contaminating contact with the white man" (1982a: 25).

All these warnings about the future of the islanders were based on the Europeans' experience of contacts between natives and newcomers with a more advanced civilization, and how ill served by these contacts the natives had been. More curiously, in the case of the South Pacific islands, a native priest of Taputapuatea by the name of Vaita foresaw the arrival of "a new kind of people coming to the island" in a widely reported trance, said to have occurred in the 1750s. Those were times of great turmoil on the island, as its fierce neighbours from Bora Bora had attacked Raiatea and chopped down some of the sacred trees. However, the new threat Vaita foresaw seemed far greater than raiding warriors.

"The glorious offspring of Te Tumu / Will come and see this forest at Taputapuatea. / Their body is different, our body is different / We are only one species only from Te Tumu. / And this land will be taken by them / The old rules will be destroyed / And sacred birds of the land and the sea / Will also arrive here, will come and lament / Over that which this lopped tree has to teach / They are coming up on a canoe without an outrigger" (Teuira Henry, 1928; Salmond, 2003:39-40).

When HMS *Dolphin* came into sight, fitting the description of a huge canoe without an outrigger, the very first one ever seen by those islanders, and thought at first to be a sacred island drifting into their midst, how could they not have been reminded of the prophecy carried in Vaitea's vision? According to Robertson, he and his messmates onboard that very first ship were regarded as demi-gods after they had demonstrated their firepower. Since thunder and lightning were associated with Oro, the god of war, there was little doubt in the islanders' mind that the new comers and their noisy guns must have been sent to punish them for past transgressions (Salmond, 2003:42).

In New Zealand, a priest by the name of Toiroa is said to have had a similar vision while in a trance. "The east coast of New Zealand / One strange people and another / Looking at each other." The vision announced the coming of Arikirangi, the high chief from the sky. Toiroa then drew in the sand a semblance of these strange people, together with their ships and the animals they would bring along (Salmond, 2003:113).

The arrival of *Dolphin*, *Endeavour*, *Boudeuse*, *Resolution*, and other ships only seemed to confirm Vaitea and Toiroa's visions. Whatever woes were to ensue were perhaps already part of the great Polynesian unconscious.

Chantal Spitz recalls the "word cast on deaf ground," warning of these comings. "They will come… these children, these branches of the same tree that gave us life… They will take our Land for themselves, they will overturn our established order, and the sacred birds of land and sea will gather to mourn." She concludes, "Everything was as it had been foretold," and the old prediction was "divinely true." Tahitians were deeply deceived when they welcomed these "strange brothers from another place."

In a faint echo of the film *Mutiny on the Bounty*, where it was reported that the sailors' white skin was their main source of attraction, Spitz writes, "We told ourselves then that their spirit must be as luminous as their skin. White, the colour of light, and therefore of intelligence. Brown, the colour of darkness, and therefore of lack of intelligence. The old prediction said 'branches of the same tree' and so that it would be right, we convinced ourselves they were the upper branches and we were the lower ones" (2007:13-14).

Prophetic native priests and Europeans thinkers uttered the same message. "This land will be taken by them. The old rules will be destroyed," predicted the Polynesian seers, anticipating the outcome. "Disease and gun powder," foresaw George Hamilton. "Crucifix in one hand and dagger in the other," warned Denis Diderot, revealing the means of the culture's future destruction. Such were indeed the gifts the Europeans brought with them when they first stopped to replenish their ships with water and fresh fruit, when they came to collect breadfruit, when they came to observe the transit of Venus, or when, like Boenachea, they brought in the first missionaries.

[VI]
THE CRUCIFIX[11]

WHEN WE READ EARLIER C.A. Lewis's description of the islands of his imagination, full of "sun-drenched sands" and "women warm in welcome," we immediately recognized the South Seas of our own imagination. Such words would immediately have alerted missionaries to the probability of a laxity of mores they would have been prompt to stamp down.

Another evocation of the islands paints a less agreeable portrait. Melville goes beyond the accepted cliché. He includes all the enchanting Polynesian details we have in mind (women, nature, outriggers), but adds a more realistic vision of its life and customs. It is a nineteenth century evocation, one that has already evolved from the one the first navigators conjured up. "The Marquesas! What strange visions of outlandish things does the very name spirit up! Naked houris – cannibal banquets – groves of cocoa-nut – coral reefs – tattooed chiefs – and bamboo temples; sunny valleys planted with bread-fruit-trees – carved canoes dancing on the flashing blue waters – savage woodlands guarded by horrible idols – heathenish rites and human sacrifices" (1982b:13).

As well as a transcultural vision of ideal feminine beauty (Arabian houri rather than Tahitian *vahine*), Melville's description now adds another layer—cannibalism, human sacrifices, "horrible idols"—to our original impression of island life spent on the sun-drenched beaches of blue lagoons, surrounded by charming and beautiful people, and always accompanied by indolent pleasures. It is the undeterred Polynesian myth. Yet, both cannibalism (to transfer the vanquished foe's *mana* into the winner's soul) and human sacrifices (to propitiate the gods) were indeed present in the rituals in these archipelagos and an intrinsic part of their culture. The reality of Polynesian life, with all its complexities, is what the missionaries met upon landing. Both the enchanting vision and the savage reality would soon meet their match.

The Missions' Impact

Missionaries were used to tackling the conversion of heathen people, a difficult task for which they followed the same principles wherever they went. They learned the language and became familiar with the local conditions and customs. They also covered the naked or semi-naked bodies, particularly those of women, forbade the pagan dances, and tried to subdue the souls. Their various denominations dictated their methods, but they all sought to shape a society closer to their own and bring on a conversion to Christianity among the native populations.

Their writings and reports also provided much valuable anthropological and linguistic information, and many modern researchers owe much of their understanding of traditional cultures, now vanished, to the observations and records left by missionaries. They and the ethnographers held somewhat similar positions, included de facto in the new way of life, yet excluded as outsiders unable to share in the culture's values. Their field narratives hold much in common, even if their objectives are radically different.

If the natives did not at first understand the missionaries' intentions and beliefs, the missionaries were no more attuned to the natives' culture. The "idols" they apparently worshipped and their beliefs were so alien to the missionaries' faith that the latter's first reaction was usually to destroy the objects, whatever artistic value they may have had, and whatever innocent rite they may have been celebrating. From the massive destruction on the Gambier Islands in 1835 of wood figurines representing various gods, only seven survived, including that of Rongo, god of the rainbow, whose everyday beneficial function was to dispense rain and ensure a good crop of breadfruit (Poignant, 1967:35). Such destructions would often make it impossible for later Tahitians to reconstruct some elements of their ancient culture and to understand the fine connections between their old beliefs, their art forms, and the structure of their society.

The missionaries' record in these matters is a depressing one. Religious missions usually followed closely in the footsteps of commercial exploitation and administrative colonization, and their influence seems fairly consistent wherever we observe them, as shown in two examples from Africa and North America, certainly indicative of a similar *modus operandi*. Piers Brendon examines the effects of colonization in Lagos, made a British protectorate in 1885. He depicts the actions of the missionaries who destroyed, perhaps unwittingly, the old religions and unravelled the interwoven fabric of society. "The British never imposed constitutional coherence… Indeed they undermined existing structures of authority. Traders assisted in this process, demoralizing the population with raw spirits that provided half the State's tax revenue. Missionaries also assisted in the process, attacking polygyny (which unified families), ancestor worship (which cemented communities), initiation ceremonies (which educated youths) and other creative inventions" (2007:527). As elsewhere, dancing was especially condemned.

In Canada, the government forbade the natives to practice an ancestral custom intended to redistribute wealth and foster social and economic reciprocity, the *potlatch*—a decision taken mostly at the instigation of the missionaries. The *potlatch* was a spiritual ceremony, present at many of life's celebrations, and sorely misunderstood by Canadian and American missionaries, who considered it a wasteful and unproductive custom, contrary to the white man's own basic interest in acquiring property. Destroying it, which the participants in the *potlatch* were seen to do, seemed a perverse nonsense to the white officials, unaware of the nature and intent of the ceremony. In some societies that seemed to promote disparities of wealth among their ranks, there existed built-in leveling mechanisms to ensure relatively egalitarian patterns of distribution, such as the *potlatch* ceremony, an intrinsic part of this social system. For the missionaries, giving away precious material goods in order to acquire intangible prestige was a foreign notion, in which they did not recognize a similar system prevalent everywhere, even in the Christian churches, because the *potlatch* manifestation only seemed outlandish to them.

Canada made the *potlatch* ceremony and the *tamanawas* dance illegal in 1884 by an amendment to the *Indian Act*. Those who infringed it were liable to a prison term of between two and six months. There was much opposition from the Indians, and some were imprisoned for continuing to practice their ancestral ceremony. However, it would prove difficult to enforce, and some officials looked the other way. It was finally repealed in 1951. The exercise had only demonstrated a misunderstanding of the role of this ceremony in the cultural make-up of a whole region. As often happens, religious proselytism and ethnocentric vision had prevailed over another culture's worldview.

This religious conflict and the legal prohibitions to which they gave rise often resulted in some customary practices going underground, such as the *potlatch* in Canada and the dances in Tahiti. Danielsson relates how the Bastille Day festivities in Tahiti took place in an orderly manner under the eye of French officialdom, then were later replaced under the cover of darkness by traditional and now forbidden dances. Gauguin described them in this manner: "There is dancing and singing, the men squatting under the trees, while the women with rhythmical movements of their arms and legs imitate the love game to which they are inciting the men, and which will begin as soon as it is night" (Gauguin, 1994:14-15). In previous times, there would have been no disrespect intended in mixing official ceremonies with dancing, since *'upa'upa* was formerly part of all popular meetings. It was banned in the name of decency and civilization from all formal functions by the missionaries, only reappearing more or less clandestinely after the white participants had left.

Most odious to the missionaries was the sight of semi-naked female bodies, which they covered with cumbersome dresses, the famous Mother Hubbards. "When they landed on the islands, the missionaries were frightened by the women's nudity and their sexuality, offered without taboo, was unbearable to them," says a Marquesan (Lefèvre, 2011). They preached decency, forbade the dances, and made the natives chant hymns in a new language. Mostly, they did their best to eradicate local beliefs. The same Marquesan explains, "The missionaries convinced their parents that their legends, their rituals were emanations from the devil and they must be abandoned and never transmitted. Thus the Marquesas became the Archipelago of Silence."

When the missionaries first arrived in Tahiti, they found clans (some no larger than extended families) usually united in similar beliefs, particularly among neighbouring islands. Their basic origin myth was similar to many others across the world, each variance appropriate to its environment. In

Polynesia, an egg that issued from primordial darkness cracked, releasing Ta'aroa (or Tangaroa). The two shells of the egg became the heavens and the earth, and Ta'aroa, the creator, made all earthly things from various parts of his own body.

Origin myths are specific to the area from which they emerge (including the scientific community's Big Bang theory). However, once past the beginning of times and the appearance of life on Earth, individual myths and the gods who accompany them all face the mysteries common to all humanity. Thus, Polynesian pantheism served to support the various myths that explained the origins of things and places,[12] as well as events of prime importance in daily life everywhere, such as the discovery of fire, the succession of night and day, the alternation of seasons, the need for agriculture, the calamity of war and the desire for peace, the distinction among the social classes, and above all the mysteries of life and death. In Tahiti, those mysteries were embodied by gods: Ta'aroa, the creator and also god of the oceans; Tane, the god of light, forests, rain, fertility; the powerful Oro or Tu, the god of war to whom sacrifices were offered and whose word was spread through the *Arioi* Society; Rongo, the peaceful god of agriculture. There were also minor gods or semi-gods, such as Maui, who gave fire to the world; Hiro, the irritating patron of thieves; another Hiro, the master canoe builder; Uahenga, who presided over tattoos; Tafai, the overseas adventurer. The names sometimes varied between islands, but they retained the same basic characteristics and functions. This pantheon of gods was in no way more primitive than the Greek one that served as classical reference in the cultural life of Europe in the eighteenth century. The names were different, and the functions changed a little, as they did in other polytheist and so-called primitive societies, but all equally served to meet people's need to interpret the natural phenomena that regulated their environment, and to give a spiritual direction to their lives. In Tahiti, all religious ceremonies merged with social activities and took place on the sacred grounds of the *marae*.

Given this religious mapping of the world, how successful were the missionaries in attempting the conversion of Tahitians and implanting their own versions of Christianity in what was at first glance a hostile and incompatible soil? From their experience in other lands, they were certainly no strangers to the pagan beliefs they found in the islands, but their successes were uneven.

The London Missionary Society

The London Missionary Society (LMS) was the first Christian body to appear on the scene after the failure of the short-lived establishment of a Spanish Catholic mission on Tahiti in 1775. LMS was no more able than the Spaniards had been to accept the natives' mores. "They hated nudity, dancing, sex (except within monogamous marriage), drunkenness, anything savoring of *dolce far niente*, self-induced penury, war (except in God's cause), heathenism in all its protean manifestations, and Roman Catholicism" (Grantan, 1963:197). Their hatred of the Catholics had little impact, but this antagonism between the Christian churches would have later consequences when the religious denominations multiplied on the islands and sometimes interfered in local politics.

Founded to spread Protestant Christianity to heathen communities, LMS arrived in Tahiti in March 1797 on board the *Duff*, a ship belonging to the Society. On that first landing, thirty volunteer missionary families, including four ordained ministers, set foot on the island. Nine other mission workers were then dropped off in Tonga; of those, three were killed almost immediately, and others barely managed to survive.

Two months later, the *Duff* dropped off two more missionaries in Tahuata, Marquesas. One left almost immediately, but the other, twenty-two-year-old Crook, managed to stay for about two years. There he struggled, unable to speak the language, his possessions soon stolen, faced with a violent culture that included eating the vanquished in battle. A year later, he was transferred to Nuka Hiva, where he lived in some comfort. He returned to London in 1799, having failed to make a single convert.

His lack of success was not the exception. Many missionaries had been unprepared and had misjudged the difficulty of their task when faced with ancient customs which included "enjoying multiple sexual partners, practicing infanticide, sacrificing and occasional human to insure good crops, performing 'lude' dances, worshipping idols, singing, and surfing" (Kirk, 2011:135-36). By 1801, after four years in Tahiti, the LMS workers had failed to convert anyone. Moreover, the loss of the LMS ship *Duff*, seized by French pirates in 1799, had delayed further expansion in this distant territory.

Crook, the young missionary so unprepared to deal with native customs, did not actually make a worse job of it than the much better established and presumably better prepared Rev. Samuel Marsden. He arrived in New Zealand in 1814, in time to celebrate Christmas, and knew what to expect. "I knew that they were cannibals—that they were a savage race, full of superstition and wholly under the Prince of Darkness—that there was only one remedy… and that was the Gospel of a Crucified Savior." It took him ten years to convert his first Maori, and another ten to convert his second (Kirk, 2011:161).

Seeing those pitiful results and the obvious reluctance of the population to be swayed by the missionaries' sermons, how can we account for the influence they were soon to display? The turning point in the fortunes of the church in Tahiti came on November 15, 1815, when King Pomare II, claiming that he fought for Jehovah, defeated his enemies and proclaimed himself a Christian ruler and king of Tahiti and Moorea. He also established Papeete as Tahiti's administrative centre.

To prove his obedience to the Christian principles preached by the missionaries and to defeat the sinful traditional customs, Pomare II enacted a number of laws against adultery, polygamy, infanticide, and human sacrifice. He also ordered people to attend Sunday church services. In both theory and practice, it was a huge step forward for the consolidation of the Christian faith in Tahiti. As members of a traditional society subject to an exacting hierarchy, Tahitians were prepared to accept obediently the king's conversion. Even if Pomare's own proclivities tended to belie his precepts, the motion had been set for his subjects to follow him in his proclaimed Christian ways.

The first real success came with the Rev. John Williams, who spent nearly a year in Moorea learning the language and the customs. He then went to Raiatea (1818), where he met with some success in converting the islanders—not so much by preaching as through sheer pragmatism. Trained as an ironmonger, he taught the islanders several useful trades, and even built a clock to encourage punctuality! In spite of some setbacks, Williams and his family persevered and eventually converted many islanders (Kirk, 2011:30-31).

The final stage of the conversion process came when King Pomare II was baptized in Tahiti in May 1819, together with about two hundred of his subjects. The same year, the missionaries wrote laws for Raiatea and Tahaa, then for Bora Bora in 1820, and for Huahine in 1822. As the islanders accepted Christianity, they also had to accept unfamiliar rules. Like so many other natives around the world, they attended church regularly, sang hymns in a barely understood language, covered their bodies in clothing, and attended school. And, as importantly, many also started forgetting their unwritten and now disparaged past.

Missionary work was often backed by local politics. In the Loyalty Islands (New Caledonia) for instance, the chiefs used the new religion to enhance their own powers. They attended church and they also imposed laws to deal with theft, adultery, and any other behaviour to which the LMS missionaries objected. It was one way of strengthening their hold over the people, who could only but follow in their footsteps. K.R. Howe, an expert on the history of the Loyalty Islands and New Zealand, writes, "There was a genuine and even aggressive enthusiasm for wearing clothes, church going, singing hymns, reciting catechisms, learning to read and write" (1988:311).

A report of the Religious Tract Society described a mass attended by an English visitor to Bora Bora in 1888.

> We found the church... holding about 500 people, a really cheerful sight. The women were all dressed in their gay-coloured sacques; their hair—which is of luxuriant growth—generally in two plaits down their back, was brightened by coconut oil... The service was given in the native language, and, after prayer, the hymn, called in the Society Islands 'Himane,' is taken up with the female voices; then the men join in with a rumbling undercurrent... After the 'Himane' came the sermon. A great rustling of pieces of paper—pieces of all kinds and colours and sizes, and pencils ditto—announced to us the fact that at Bora-Bora, notes were taken and it was accounted for in this way, that the congregation are examined on the subject of the sermon during the following week, and the notes are carefully preserved for references.

The dual rivalry between the Protestant English and the Catholic French interests in Polynesia are also alluded to in the same report.

> We were assured by many that the reason why a foreign [i.e. French] protectorate had been declined, was because they all felt assured that the happy state of the islands was solely attributable to the English missionaries who had brought the Bible to them, and had, by their own example in everyday life, shown them practically how to carry out its teaching to their own happiness and the peace of their neighbours. Their influence had certainly wrought wonders among the natives.... In Bora Bora at 8 p.m. a curfew bell is rung for the people to return home, and at 9 p.m. a second, after which no one is allowed out. This is done every day. On Sunday, no trading is allowed, no canoe leaves the shore, no fishing canoe comes off, and when you land all is quiet, peaceful, and happy, the cheerful prattle of the little children giving life to the scene. No spirits or strong drinks are allowed on the island, and a heavy fine is inflicted on the any native found possessing any, while the fine for a chief found drunk is double. Now this state of things would not last long under a French Protectorate, judging from what exists in other places (1888:11-13).

In spite of the islanders' natural docility and occasional good will, the Europeans had often been mistrusted in the past. Consequently, LMS-trained Polynesian preachers were sent by the 1820s to pave the way for the European missionaries arriving later. This proved to be a successful practice, particularly in Rarotonga and Aitutaki. When two missionaries from LMS arrived in Rarotonga in 1827, they were astonished to observe how efficient the local Raiatean missionary had been in converting the population. He had in fact exceeded his mandate by establishing a morality police, called *rikos,* who actively searched the island for any breach of religious precepts. The visiting Rev. Pitman noted that the *rikos* had "usurped so much authority that they were objects of fear more than love." He was upset to discover that they had obtained compliance by using a gun and that they had also acquired some thirty "pieces of property" (Kirk, 2011:203).

The influence of the *rikos* was well known on Rarotonga, with as many as one out of every six people recruited. They reported on any aspect of the islanders' behaviour deemed offensive to the Church, including working on the Lord's day, being affectionate in public, drinking *kava,* or being out after nightfall. There were ways to prevent such misdeeds and forestall untoward behaviour, particularly with a member of the opposite sex. For example, when holding his sweetheart's hand in public after dark, a man was supposed to keep a lit candle in his free hand so he could not embrace her (Kirk, 2011:206).

From Rarotonga also came news of a model Christian village established at Avarua by visiting LMS missionaries. These missionaries were to serve there without interruption or evident opposition during some thirty years, until 1860. When the inhabitants, who had never before lived in close proximity to one another, were pressed to move into a village, they suffered a major alteration in their lifestyle. But living together made it also much easier for the missionaries to monitor their activities and to enforce the application of the rules they had created.

As would happen on several islands, the villagers of Rarotonga were put to work and built a church made of coral blocks. Unlike the attractive Catholic cathedral and other churches on the Gambier Islands (albeit built at such heavy cost and with such disastrous consequences), a certain Rev. Buzacott's construction found few admirers, and the Earl of Pembroke called it "that vile, black and white abomination paralyzing one of the most beautiful bits of scenery in the world" (Kirk, 2011:206).

In 1826, members of the LMS in Tahiti realized that a new faith was making inroads into the converts' religious practices, and that a new current was emerging. The new sect combined elements of Christian theology with ancient Polynesian beliefs, and was called *mamaia*. Local prophets claimed to have heard the voices of Christ and the saints, and preached the early coming of Christ, while also accepting the millennial prophecies then enjoying a revival. The Bible served to provide examples that were used to draw parallels between the prior Tahitian sexual freedom and King Solomon's harem. Although the missionaries expressed deep concern, the *mamaia* cult spread to Raiatea, Tahaa, and Bora Bora (Rowe, 1988:151; Kirk, 2011:199-200). The millennium movement was to see its darkest hour and demise in 1841, after competing for a decade and a half with the more traditional Christian church. The devastating smallpox epidemic in Tahiti dealt the *mamaia* cult a fatal blow, as its adherents had refused to be inoculated and died in large numbers. Although lasting only fifteen years, the movement had an impact on the islanders, who were still struggling to accommodate their traditional beliefs within the new religious doctrines they were being taught.

Churches run by LMS shut down on Tahiti after the French takeover and the English missionaries left Tahiti in 1886 and the other Society Islands in 1890. However, a member of the *Société des Missions Evangéliques*, French pastor Charles Vienot, took over the responsibility for protestant churches, missions, and schools. Today, the Protestants outnumber the Catholics in Tahiti.

The Catholic Picpusians

The Spanish Friars who had come briefly with Boenachea in 1775 had seemed very alien to the Tahitians, with their incomprehensible refusal to fall prey to the native women's charms. "They all unanimously condemn'd the flesh-subduing Dons, for that self denial which may be deemed meritorious in Cells & Cloisters, will be always looked upon with Contempt by the lovely & beautiful Nymphs of Otaheite," wrote David Samwell, one of Cook's companions (Salmond, 2009:423).

When the Catholics came back to Tahiti, from France this time, the main area of conflict once more related to sex. Salmond outlines the deep conflict of values that opposed Tahitians and the Catholic missionaries: premarital sex vs. chastity, and polygamy and trial marriage vs. monogamy (2009:240). We can imagine how foreign the new religious concepts must have appeared to the natives when forcibly introduced. The arrival of the French missionaries was the beginning of a more intensive confrontation between the islanders and the emissaries from the Church of Rome.

Three priests arrived on the Gambier Islands in 1834. They belonged to the Order of the Sacred Hearts of Jesus and Mary and of the Perpetual Adoration of the Blessed Sacrament of the Altar—otherwise known as the Picpusian Fathers after the address of their headquarters, rue Picpus in Paris. When they arrived in Mangareva, they found an agricultural population growing breadfruit, taro, and yams, and fishing offshore. In spite of this apparently idyllic lifestyle, warfare was endemic among the tribes. The winners would systematically destroy the food reserves of the losers, who would then seek revenge in an endless cycle of retaliation.

As in most Polynesian islands, the high chiefs had total control over their people. The aristocratic *ariki* controlled the land, while the commoners worked it. So, when the Picpusian fathers succeeded in converting the chiefs, who often found such a conversion to be to their advantage, the commoners had no choice but to follow suit. It took little time for the fathers to order the destruction of the pagan statues on the islanders' sacred *marae* and to ban singing and dancing, then at the heart of all ceremonies. The habit of pious obedience was such among the islanders that they presented no objection to being converted. On Mangavera, one priest claimed to have baptized 1,568 people in four years.

The first missionaries' goal was to convert; later, once settled and at the head of a congregation, they worked at providing medical care and educating the natives. The French Catholic missionaries were reputed for the large number of well-built religious and civil edifices they erected on the Gambier Islands between 1840 and 1880. One individual in particular, Father Laval, became known for his ambitious building plans. Among his first projects was the Cathédrale St Michel, a building of white coral rocks decorated with mother-of-pearl, and the largest church in the Pacific at the time. Another church, Notre-Dame de la Paix, conceived as a miniature Cathédrale de Chartes, was also built on Aukamaru in the same exhausting conditions by carving out and retrieving coral stones from the reefs, and moving them to shore on heavy rafts.

By the time the Cathédrale St Michel was finished in 1848, some five thousand Gambier islanders had died of overwork, disease, and starvation. Starvation seemed ironic in a land where nature is so generous, but taking men away from their customary agricultural work and fishing activities had a radical effect on the resources available to the community, and many people became severely undernourished and even starved. Slave workers had to be brought in from the Tuamotous (either abducted or enticed to come to Mangareva) to substitute for the declining population. Upon hearing the number of deaths, Father Laval is reported to have said, "True, they are all dead. But they have gone to Heaven the more quickly" (Kirk, 2011:247). He completed his ambitious building program in several parts of the Gambier Islands, leaving a large number of churches, presbyteries, convents, schools, and lighthouses.

Father Laval was to rule unchecked over the bodies and souls of the Mangarevans for three more decades until 1871, when the French authorities were finally alerted to the terrible loss of life on the Gambiers. The fall of Napoléon III in Paris the year before had made way for an anticlerical government in France, and the Picpusian fathers no longer had the same influence at home. Father Laval was relieved of his duties and brought to Tahiti, where he faced a charge of murder. However, he was judged to be insane, escaped trial, and died soon afterwards.

Four years after their arrival on the Gambiers, the Picpusian priests came to the Marquesas on board the *Venus* (1838). These were cannibal islands with a fierce topography, and previous missionaries had failed to convert anyone. But the new Picpusians, better prepared than their predecessors, knew the language and the customs, and started befriending the chiefs in a system of traditional name exchanges that ensured trust and secured reciprocity of services, as had the navigators before them. Moreover, the islanders had not yet learned to distrust and resent them as they did the other whites whom they had previously met with unfortunate consequences (Kirk, 2011:243).

When France annexed the Marquesas Islands in 1842, a few of the islanders offered little resistance because the Fathers had gained their confidence. Considering that the priests had been actively destroying Marquesan culture by the usual means (proscribing cannibalism and infanticide, but also—as if these carried the same importance—traditional nudity, dancing, singing, promiscuity, polygamy, and tattoos) the islanders' trust could seem somewhat surprising. During the first decade of their presence in the Marquesas, the Picpusian Fathers claimed to have baptized 216 people (Kirk, 2011:240, 262).

There was constant strife between the two sets of Christian missionaries and their competition was equally constant. At the request of the Protestants, Tahiti expulsed the French priests in 1836-37. France immediately demanded reparations and even threatened to destroy Tahiti unless a ransom was paid and Queen Pomare issued an apology. Further, they requested that French priests be made welcome from then on. The Queen had attempted to obtain British protection against the French threat, but such a move was not part of the British policy of the day.

Melville confirms the antagonism between the two churches and the violent reactions to the French efforts to establish a Catholic mission in Tahiti. In one instance he wrote, "Two priests, after enduring a series of persecutions, were set upon by the natives, maltreated, and finally carried aboard a small trading schooner, which eventually put them ashore at Wallis—a savage place—some two thousand miles to the westward." He also reports that the English missionaries had acknowledged having a part in these persecutions since, by "their inflammatory harangues they instigated the riots which preceded the sailing of the schooner" (1982b:450).

The same conflict would continue as long as the missions were actively working at converting people and establishing a following. Like Melville, Gauguin was to witness the competition between the two sets of Christians, although in his case it took on a lighter form. When asked to judge singing competitions, he had not anticipated that the rivalry between the Catholic and the Protestant missions would become evident during these festivities. Yet, "each mission sent a party of dancers from amongst their converts to compete for the prize, and the only way I was able to satisfy them both (I say "I", because my committee were all so jealous of each other, and also so very far from sober, that they would only accept a decision from me, a stranger) was by giving each faction a first prize, and even then the two ascetic-looking Jesuit priests regarded me with far from friendly glances" (Danielsson, 1965:255).

The Case for the Missionaries

The establishment and progress of the Christian faith in Polynesia varied among the different islands and archipelagos. At different times, with different churches (LMS and Picpusians), and with different means at the missionaries' disposal, the burden of the conversion and sustenance of the faith was experienced differently by the islanders.

Living with the emissaries of Christ in Moorea was different from living with them in Rarotonga and in Mangareva. As to the Hawaiian Islands, they were in a different situation as the only Polynesian culture in which the internal abolition of some *kapus* (taboos) and the destruction of some temples caused the traditional religion to collapse even before the arrival of the missionaries (Kirk, 2011:172). Everywhere in Polynesia, the imported religion was the determining factor in destroying the existing culture.

John Gilbert (1973:313) makes the case that there were good and bad sides to the arrival of Christian missionaries in the Pacific islands. He deplores the "substitution of an alien faith for a long established pattern of beliefs and customs." It replaced the original warmth and meaning their own beliefs and customs brought to the inhabitants with the boredom and bewilderment they experienced when faced with the new and imposed simulacrum of a religion (notwithstanding the fact that there were indeed some true native converts to Christianity). The imposition of a foreign faith on a traditional population has had devastating effects, not only in Polynesia but arguably everywhere Christian missionary work has accompanied European explorations and colonization.

Against these negative effects, Gilbert opposes the positive side of their often successful battles against drunkenness, prostitution, and slave labour. However, since these were actually brought into Polynesia by the white men themselves, it is difficult to feel much gratitude towards the missionaries for trying to control them. It may seem unfair to bring together the often well meaning missionaries and the harsh whalers and the corrupt slavers, yet they all belonged to the same dominant and exploitative society that succeeded in transforming Tahitian culture.

Gilbert adds that the missionaries' presence also contributed to the eradication of previous primitive practices, not uncommon across Polynesia, including cannibalism, ritual sacrifice, and the killing of newborn infants. These occasional ritual or social practices were an intrinsic part of the culture, and they disturbed the first Europeans who discovered them and the later travellers who also encountered them. None of these observers, with their ethnocentric view, could have possibly understood these practices or accepted them. No more than the islanders could possibly have understood the Europeans' revulsion at seeing these practices performed. It is a truth now commonly acknowledged that ethnocentricity works both ways.

Gilbert's final argument is that the preparatory work done by the missionaries and nuns eventually made possible the establishment of schools, hospitals, and other welfare services provided under the colonial rule in the

1900s and now brought to their current levels of efficiency. It would be difficult to counter these arguments, yet for those modern islanders who reject what European "progress" has brought them, the argument may be weaker than Gilbert appears to think. The fact that epidemics are now more easily controlled thanks to modern European medicine does not detract from the fact that these epidemics were themselves European imports.

There are many testimonies to the efficacy of the missionaries' methods and to the improvement brought to the islanders' previously loose morals and warlike proclivities. Among them, Duperrey, a hydrologist who mapped the Gilbert Islands archipelago and who commanded *La Coquille* during a French circumnavigation (1822-25) wrote in 1823, while berthed in Tahiti, that idolatry no longer existed and the missionaries had totally changed the morals and customs of the inhabitants. He could only see virtuous and reserved women everywhere, and believed that the former fierce wars and human sacrifices were things of the past.

Otto von Kotzebue, the early nineteenth-century navigator and explorer who also examined the contribution of the missionaries to the islanders' culture, was less forgiving than Gilbert or Duperrey, and his assessment was scathing. "The religion of the missionaries has, with a great deal of evil, effected some good. It has retained the vices of theft and incontinence; but it has given birth to ignorance, hypocrisy, and a hatred of all other modes of faith which was once foreign to the open and benevolent character of the Tahitians." Cameron, who quotes him, adds a more modern interpretation of what has happened to the culture: "Even worse than this, it has left them quite defenceless, incapable of combating the most insidious of all emotions—boredom, a profound boredom bred of the congenital indolence of the South seas" (Cameron, 1964:53).

Gilbert expresses a more general approval of the missionaries' work than did Kotzebue. Darwin, who paid a brief visit to Tahiti in 1835 *(A Naturalist' Voyage,* Chapter XVIII) and was convinced of the "morality and religion of the inhabitants," also subscribed to the belief that the natives were better off under the influence of the missionaries. First, he said, that having "reflected on the effect of intemperance on the aborigines of the two Americas I think it will be acknowledged that every well-wisher of Tahiti owes no common debt of gratitude to the missionaries." Naturally. Yet we should not forget that the native *ava* or *kava*'s inebriating virtues were in no way comparable to the dreadful ravages of alcoholism that spread through the islands as soon as alcohol brought in by the Europeans could be easily obtained by all.

Next, Darwin brought back the arguments that Tahiti's ancient culture was essentially evil and redeemed only by the new adherence to Christian principles. Thinking mostly of Kotzebue among those who attacked the missionaries, he wrote, "They forget that human sacrifice, and the power of an idolatrous priesthood—a system of profligacy unparalleled in any other part of world—infanticide a consequence of that system—bloody wars, where the conquerors spared neither women nor children—that all these have been abolished and that dishonesty, intemperance, and licentiousness have been greatly reduced by the introduction of Christianity" (Chapter XVIII).

I am not an apologist for the good old Tahitian ways, but I must remark again that outsiders who attempt to transform another society to have it conform to their own beliefs and standards for religious and commercial purposes, and do so quickly, probably do irreparable damage to that society. Even if, like Phoenix, it could eventually rise out of its ashes, it would be transformed almost beyond recognition, having lost most of its connections to its past, good and bad.

The missionaries' task was made more difficult by the presence of transient white men, mostly traders and whalers, who constantly undermined the Church's precepts by bringing alcohol to the islands and encouraging prostitution. One such example of this frustrating behaviour occurred in Hawaii, where the early missionaries and the rowdy whalers were always on opposite sides. In Lanai, then the centre of both the whaling industry and prostitution, crew members often came ashore with only two purposes: to get drunk and to have sex. The New England Congregationists on the island were obviously loath to see such activities and did their best to stem the stream of drunken, brawling, and cavorting sailors in their streets. One famous incident, in October 1825, could have turned very violent. The crew of the whaling vessel *Daniel* roamed the streets of Lanai for four days, menacing all those they suspected to have had a hand in preventing them from coming ashore. They even attempted to hang Rev. Richards, whom they blamed (probably rightly) for the opposition. He was only saved when his wife, leading a group of Maui Christians, chased the sailors back to the *Daniel* (Daws, 1968:93). They left, taking along the women already on board.

More insidious, but no less frustrating, were some Europeans' actions intended to subvert the missionaries' efforts to regulate their congregation's life. Perhaps not unexpectedly, Gauguin was often in conflict with the Catholic Church and tried to undermine its influence, both on general principle and through self-interest. For instance, he told parents that their thirteen-year-old daughters did not have to go to school. At stake was the age of the girls, whom missionaries and teachers considered to be still young enough to attend school, while Gauguin, whose own *vahine* had seldom been older than fifteen when he first met them, reclassified the same girls as women.[13] It could be argued that his attempts were not merely self-serving but also aimed at reinstating pre-missionary conditions and mores and, thus, beneficial to a culture that had been deprived of its customs. However, times had changed and a century of European influence could not be eradicated by simply declaring that a thirteen-year old female was no longer a child and could become a "wifie."

Today, if we look at Papeete from the sea, we see clearly that the city's commercial core seems anchored by two churches on opposite ends of town: the Cathedral to the left on Place Notre-Dame, and the Protestant Temple to the right on rue du Temple. They seem a little lost within the modern buildings and shopping areas, but their presence is both noticeable and symbolically polarizing, with an established history of rivalry between them.

The two Churches have a strong following in French Polynesia. The Protestant missions were the first to be established (if we ignore the futile attempt of the early Spanish missionaries) and today's followers constitute about forty-five percent of the population, showing particularly strongly in the Austral and Leeward Islands. The influence of the Protestant Church was bolstered by its involvement in social matters, such as opposition to the French nuclear testing.

Settling next, the Catholics still have a strong following, particularly in the Tuamotus and the Marquesas. They are evenly split between Polynesians and *demi* (descendants of Europeans who fathered children with Polynesian women).

Naturally, these two faiths are not the only ones fighting for the souls of the islanders. The first Mormon missionaries arrived in 1844 in Tubuai (Austral Islands), and some ten percent of the population still follow their doctrine. From the beginning, a small breakaway group from the traditional Mormon faith developed in the Tuamotus, the Reorganized Church of the Latter Days Saints, who reject Brigham Young as a second prophet. They are known as the Sanitos.

Finally, the Seventh-Day Adventists constitute up to ten percent of the population and the Chinese Buddhists, about five percent. The Chinese *(tinito)* are descendants of those brought in to work on the cotton fields in the 1860s, later becoming small farmers and shopkeepers

One may wonder how modern Tahitians deal with the knowledge of their recent and often appalling history and the damage the missionaries did to their culture, while remaining faithful to the new religions they have been inculcated. Attendance in the various churches and temples show them to be religious people, which is not surprising since their ancient culture was wholly based on close connections with their gods, and religious rituals were part of their daily lives. Their *marae* were meeting places for all spiritual, social, and political activities.

If we consider the case of Mangareva, and the physical and moral cost of the missionaries' activities there, we can better appreciate the manner in which some islanders have been able to reconcile the opposing forces of their religious faith and the human failings of those who imposed it in the name of their God. Modern Mangarevans know their past and probably do not cherish the memory of the Picpusians. Nevertheless, they are united in the *Association Sauvons la Cathédrale* to restore their beautiful old cathedral, built at such human cost. Great efforts are being made to rebuild it using old materials and tools, and relearning ancient methods and craftsmanship. Their attitude appears to combine their religious spirit and a strong desire to renew with the traditions of their past (Lenglant, 2010).

On the whole, the islanders seem to have received and absorbed the Christian messages of piety, obedience, and compassion. They also seem to have developed a healthy means of distancing themselves from the often harsh ways the missionaries sought to introduce those virtues. With the passage of time, they also seem able to reconcile these new faiths with some aspect of the old beliefs that connected them to nature and were an integral part of their traditional culture.

The *mamaia* movement, although now vanished because their adherents died of disease, and the *sanitos* in the Tuamotous, who adapted the Mormon message to suit their needs, may indicate that there are possible compromises to be made with established churches. The large array of the latter in Tahiti, their proselytism, and the following they generate show the people's willingness to participate in religious rituals. They probably also indicate a spiritual flexibility among the islanders that may lead to some further dogmatic deviance beside the two already noted.

Christianity may well be forced to compromise and accept that more traditional beliefs could blend comfortably with its own creed. As we read Spitz and Vaite's novels, we see occurring a workable mixture of Catholicism and former animism, the two sets of beliefs seemingly able to coexist. Spitz writes approvingly of the double protection offered by "the spirits of the Fathers who still watch over the village" and by "the ministers' God, who is there on Sunday, the Lord's day" (2007:23).

Vaite's characters experience a similar compromise. Loana, who is a good catholic, knows and probably believes the Tahitian creation myth. But she also has in her house a statue of the Virgin Mary ("Understanding Woman") and says, "It does not mean that the story about Adam and Eve is an invention."

There does not seem to be a significant gap between Materena's great-grandmother who said to a priest, "You come here and burn our prayer-meeting houses, and destroy our *marea,* you say our God does not exist—your God you can keep for yourself, I don't want him," and her mother who, asked if God really exists, "did not say yes, she did not say no. She just said, 'One day, girl, you're going to be thankful there's a god for you to believe in.'" The key to this ambiguity is found in Loana's explanation. "We, the Polynesians, have always been a religious type of people. In the old days [a long time ago], we prayed before everything we did. We prayed before eating, working on the land, planting our gardens, throwing the net, and before we began and ended a voyage. We prayed non-stop" (2006:211-13). This religious predestination, which seems so deeply implanted in the culture, can probably accommodate itself to various types of worship.

This adaptability could indeed be verified through the renaming in 2004 of L'Eglise Evangélique de Polynésie to l'Eglise Protestante Maori. Unlike the Catholic and Mormon churches, this Protestant church attempts to reconcile Christianity with the Polynesian cultural heritage, and reinstates tattooing, the *haka,* the *heiva,* and a renewed interest for the *marae* of the past. However, it is difficult to assess to what extent their former cultural significance has been preserved in this modern version.

[VII]
THE DISEASES

THE EARLIEST AND MOST EVIDENT pathogenic evil brought in by the navigators' crews was the spreading of venereal diseases and tuberculosis, known respectively at the time as "Cupid's itch" or "the curse of Venus" and as the "English disease."

Everyone had observed how well made and healthy the inhabitants of these pristine islands appeared to be, when first meeting them in the eighteenth century. However, there were, and there still are, a number of endemic diseases, such as yaws and elephantiasis. The latter is well established on the island and so easily acquired that Danielsson and his wife were both contaminated in the 1950s after only a few months and had to leave Hivaoa, in the Marquesas, to recover.

Yaws, a chronic condition caused by the spiral-shaped spirochete bacterium, was endemic at the time in the Pacific. Unlike syphilis, it is not transmitted sexually but by direct contact with the skin of infected people. The similarity of several symptoms between the two diseases led to confusion and to the suspicion that syphilis already existed on the island before the arrival of the Europeans. While syphilis rapidly spread among the islanders, none of the Europeans seem to have contracted yaws.

Elephantiasis is caused by tiny larvae circulating in the blood, resulting in a swelling of the arms, legs, and genital organs. It is transmitted through mosquito bites. Danielsson (1957:98) mentions that the two deterrents to the spreading of the disease, eradicating the mosquitoes or isolating the sick persons, were not at the time considered valid options by the islanders—they thought hunting mosquitoes was silly and isolating people was unkind. They did not believe the white men's theories of how the disease is carried, believing instead that this disease was caused by certain kinds of food.

When Herman Melville visited the island, *Fa-Fa*, or elephantiasis, seemed prevalent. Affecting the legs and feet alone, the disease swelled the latter "to the girth of a man's body, covering the skin with scales." He commented, "It might be supposed that one thus afflicted would be incapable of walking; but to all appearance, they seem to be nearly as active as anybody; apparently suffering no pain, and bearing the calamity with a degree of cheerfulness truly marvelous" (1982b:455). No such leniency would be afforded the islanders after the arrival of the truly dreadful diseases that were to afflict them after their contact with the Europeans.

Polynesians had terms for leprosy, bronchitis, abscesses, and impetigo, but none for the conditions that were to devastate their island, particularly the terrible epidemics that would decimate their numbers, including typhoid, influenza, smallpox, and venereal diseases. Unlike the people of Europe, Asia, and Africa, the Polynesians had lived in such isolation that had never had the opportunity to develop any immunity to the diseases that were to ravage their population and transform their culture.

The Pox

It is widely acknowledged that the Europeans introduced venereal diseases into Polynesia. Specifically who of the British, Spaniards, or French were responsible is probably impossible to determine with accuracy, even though blame was attributed to each in turn. A succession, even a convergence at times, of ships from all three nations anchoring there within a short period must have made it difficult to pinpoint the source of the infection, even if some captains were not shy about pointing the finger at their foreign colleagues.

For Wallis, there was no doubt that the French were responsible. He had already been riled at the unforeseen trading of the women's favours for the nails his sailors stole from the ship (this trading seemed somehow to have escaped Bougainville's notice, but it seems unlikely that the charms of French sailors so far outshone those of the British that they did not have to parlay similar trades themselves). Then, Wallis found himself having to defend the honour of his country against the rival French in a field where neither would have wished to compete. He first established his men's state of health. "It is certain that none of our people contracted the venereal disease here, and therefore, as they had free commerce with great numbers of the women, there is the greatest probability that it was not then known in the country." Yet it had been discovered by Cook, and the obvious conclusion was to look for a ship that had landed between their two visits. "As no European vessel is known to have visited this island before Captain Cook's arrival but the *Dolphin,* and the *Boudeuse* and *Etoile,* commanded by M. de Bougainville, the reproach of having contaminated with that dreadful pest a race of happy people to whom its miseries had till then been unknown must be due either to him or to me, to England or to France, and I think myself happy to be able to exculpate myself and my country beyond the possibility of a doubt" (Hanbury-Tenison, 2005:408).

The disease had obviously been brought in by the Europeans, but why should Wallis have thought his English crew innocent of spreading the infection while the French one bore the sole responsibility for it? Bougainville's men on *Etoile* and *Boudeuse* very likely suffered from both scurvy and venereal disease, and it is just as likely that the same ailments would have afflicted those on the *Dolphin,* so Wallis's righteous defense of his men will probably never be validated. In fact, some authors are not shy of attributing to Wallis's crew the responsibility he so strongly denied. "On July 27, when *Dolphin* departed, Tahitians wept. Tahitians would continue to weep: Crew members had infected island females with venereal disease. Sexually transmitted diseases were a European import and had not occurred in Polynesia before contact with white sailors" (Kirk, 2011:49).

The first to arrive in 1767, Wallis may indeed have felt that he had taken the necessary precautions by having the ship's surgeon check every man on board before allowing him to go ashore. They all seemed to be free of any sign of the disease.

Bougainville had been no less careful than Wallis. He remarked on the natives' health and his concern not to introduce new diseases to the population. "Our surgeon assured me that he had on several of them observed marks of the smallpox; and I took all possible measures to prevent our people's communicating the other sort to them; as I could not suppose that they were infected with it" (Hanbury-Tenison, 2005:412).

Similarly, before landing in Tahiti, Cook had ordered Dr. Monkhouse to check the crew for gonorrhea and syphilis. According to the surgeon, only one man was infected. Cook confirmed that he "had taken the greatest pains to discover if any of the Ships Company had the disorder upon him for above a month before our arrival here and ordered the Surgeon to examine every man the least suspected who declar'd to me that only one man in the Ship was the least affected with it and his complaint was a carious shin bone; this man has not had connection with one woman in the Island" (Beaglehole, 1974:128).

While the ships' captains may have acted responsibly, given the medical knowledge at the time, this may not always have been the case with those on whom they relied, the ships' surgeons. Some of the *Dolphin*'s men bore visible signs of infection that could not have escaped the surgeon's attention. Salmond quotes George Robertson, the ship's master. "We carried Ten Men ashoar [at Tahiti] to the Sick Tent, three of them Very Bad, and has Been so Ever since we Left England, with Damn'd veterate Poxes and Claps" (2003:52).

There were obvious signs of venereal diseases in the island during Cook's visit. By early May, although none of the *Dolphin*'s men had contracted the disease while on the island, several now exhibited its symptoms. Cook wrote that he had "reason (notwithstanding the improbability of the thing) to think that we had brought it along with us which gave me no small uneasiness and did all in my power to prevent its progress, but all I could do was of little purpose for I may safely say that I was not assisted by any one person on the Ship… this distemper very soon spread it self over the greatest part of the Ships Compney but now I have the satisfaction to find that the Natives all agree that we did not bring it here" (Beaglehole, 1974:188).

So, it would appear that the three captains involved within that very short period of time, Wallis, Bougainville, and Cook had taken whatever precautions they thought necessary to preserve the health of these innocent islanders. However, Beaglehole comments in his *Life of Captain Cook* that, according to the ship's records, one third of the ship's company were affected. Little was known at the time of the disease's varieties and latent periods, so Cook felt satisfied that neither the *Dolphin* nor the *Endeavour* were responsible for the spreading of the disease in Tahiti (1974:187-88).

Bougainville shared the same belief that his own crews had not been responsible for introducing the disease in Tahiti. Indeed, it was a shock for Commerson, a surgeon as well as a naturalist, to discover cases of venereal diseases among the *Etoile*'s men after leaving Tahiti. He immediately assumed they had been contaminated by the Tahitian women and, believing in the innocence of the natives, thought that Wallis's men had been the ones to introduce gonorrhea and syphilis to the island.

We can say with some certainty that both English and French captains had done what they could to fulfill their responsibilities towards the islanders and to prevent spreading the disease to them. Doing their best was obviously not enough, and Salmond notes (2009:162) that at least seven of the British crew had been treated for venereal diseases during the month before they reached Tahiti and anyone who had contracted either gonorrhea or syphilis during their visits to Rio de Janeiro and Tierra del Fuego would have still been infectious, even if they no longer presented any symptoms.

When some of the sailors, previously cleared by the surgeon, started showing symptoms, Cook was bitterly disappointed but found little cooperation from officers, sailors, or even members of the Royal Society who, as scientists, should have known better. In spite of Cook's precautions, it was a lost cause. He reported, "I was oblige'd to have the most part of the Ships Company a Shore every day to work upon the Fort and a Strong guard every night and the Women were so very liberal with their favours, or else Nails, Shirts & were temptations that they could not withstand" (Salmond, 2009:153). Still, Cook persisted in his efforts, and his log entry on *Resolution* for 25 January 1779 carries the information that "W. Bradley, Seaman, was given 24 lashes for having connections with a native woman knowing himself to have the venereal disorder."

By the time of Bligh's arrival, venereal diseases were definitely to be found everywhere on the island. Salmond comments that during Bligh's visit to Tahiti, "more than a third of the *Bounty*'s crew received treatment for venereal infections [the same ratio as on Cook's *Bounty*]... Since syphilis, for instance, can be latent for long periods, with no visible symptoms, it was often not diagnosed during the 18th century and it is possible that other members of the crew were already infected with this malady when they arrived at Tahiti" (2009:163). Those who had not yet shown signs of infection before their arrival very likely contracted the disease during their time there.

These diseases had definitely been introduced by Europeans, even if neither the British nor the French felt any particular responsibility for doing so. Could there have been another source of contamination? A Tahitian chief told Cook about two other ships that had visited the island earlier, and that it was believed they had "brought the Venereal distemper to this Island where it is now as common as in any part of the world and which the people bear with as little concern as if they been accustomed to it for ages past" (Beaglehole, 1974: 188). From the description, the affliction was thought to be gonorrhea rather than syphilis. Cook showed the chief different flags and the latter at once picked out the Spanish flag as the one flown by these earlier vessels. As well, based on some European garments he also recognized, Cook believed beyond doubt that the ships were Spanish from some South American port[14] (Beaglehole, 1974:189). Salmond, on the other hand, thinks the reference was to Bougainville's ships (2009:164). It matters little today which of the European ships brought in the disease, as none were exempt from carrying it on board, but it would have eased the captains' conscience at the time to determine that it was another nation's responsibility for carrying this burden to an archipelago most of them had enjoyed and some had even cherished.

Although the Tahitians knew this new disease had been introduced by the foreign visitors, they did not seem to have made the connection with the manner of its transmission. Bowes Smyth, surgeon on the *Lady Penrhyn*, wrote that eleven years after the 1777 visit of the *Resolution* "although many people had died from venereal diseases … the sailors and women slept together… They [the women] were total strangers to every idea of shame in their Amours" (Salmond, 2011:136).

It was common in many traditional cultures to associate whatever evil befell them to causes that our modern scientific minds would consider to be superstitions, such the "evil eye." As to the Tahitians, they thought these diseases "originated in the displeasure of some offended deity or were inflicted in answer to the prayer of some malignant enemy" (Salmond, 2003:452). The gods' anger against those who might have transgressed sacred restrictions (*tapu*) was sufficient to wreak havoc among the tribes.

THE NEW ARCADIA: TAHITI'S CURSED MYTH

Eventually, no island would be spared, from Tahiti to Queen Charlotte Sound in New Zealand, where Captain Furneaux's infected crew on the *Adventure* repeated the familiar pattern: nails for sex with native women, with the unavoidable consequences.

The missionary John Williams confirms that the natives believed that all the diseases he observed on the islands he visited had been introduced by ships. It reveals once more the banality of the situation and the complete lack of intent for the ensuing devastation. "What renders this fact remarkable is, that there might be no appearance of disease among the crew of the ship that conveyed this destructive importation, and that the infection was not communicated by any criminal conduct on the part of the crew, but by the common contact of ordinary intercourse" (Williams, 1837:281-82).

Epidemics

The Europeans' diseases followed them wherever they went. In most cases, the native populations had no immunity to them. The devastation we will see in Tahiti had previously exploded in the Americas with equal virulence. Fernand Braudel relates how viruses, bacteria, and parasites were imported to the New World from Europe or Africa.

> The Europeans had hardly set foot in the New World before smallpox broke out in Santo Domingo in 1493; it appeared in 1519 in besieged Mexico City, even before Cortez reached it, and in Peru in the 1530s, before the arrival of the Spanish soldiers. It spread to Brazil in 1560 and to Canada in 1635. This disease, against which Europe had become partially immunized, made deep inroads into the native populations. The same was true of measles, influenza, dysentery, leprosy, plague (the first rats are said to have reached America in 1544-6), venereal diseases,… typhoid and elephantiasis. All these diseases, whether carried by whites or blacks, took on a new virulence (1982:37).

I would be remiss in not mentioning my own part of the world, British Columbia, where the catastrophic smallpox epidemic of 1862 reduced the northern native populations in the most heartbreaking way. The Tsimshian lost about half of their people, the Kwakuitl two-thirds, and the even more unfortunate Haida three-quarters.

Knowing this pattern, the events in Polynesia in the eighteenth century should hold no surprise for us. George Hamilton, surgeon on the *Pandora*, commented on another disease also to be found among Polynesians which, he felt, had done as much harm as the venereal diseases now prevalent everywhere, "a disease of the consumptive kind has of late made great havoc amongst them; this they call the British disease, as they have only had it since their intercourse with the English" (Salmond, 2011:297).

Few contemporary observers had noted the emergence among the islanders of tuberculosis, another European import. At first, it seemed to have affected crew members more seriously than the Tahitians themselves. The disease apparently spread from ship to ship as sailors moved around and contaminated others. The novelist Fanny Burney (well acquainted with naval news through her brother James Burney, who sailed with Cook on two voyages), upon hearing that John King had died of tuberculosis contracted from two of Cook's officers, wrote, "'Tis very strange that all the circumnavigators though they seemed well at first, are all now apparently broken in their constitution. There are now eight others falling into the same premature decay" (Salmond, 2011:109).

There may not have been references made at the time to the spreading of tuberculosis among the natives, but LMS missionaries were later to report that the islanders had blamed the English for having introduced many virulent diseases to their island. "They say, that captn. Cook brought the intermitting fever, the humpbacks & the scrofula which breaks out in running sores in their necks, breasts, groins and armpits" (Salmond, 2003:454). The latter may have been tubercular lesions.

However, venereal diseases and tuberculosis were only the beginning. As whalers, traders, and planters arrived, more numerous all the time, they also brought other diseases from which the natives had neither immunity nor means of control. The lack of resistance of native populations was already known by then. When Mai was taken to England on the *Adventure* to meet the British monarch, the latter recommended to Joseph Banks, who was presiding at the introduction, that Mai be inoculated against smallpox so that he would not succumb to the disease like other "savages" who had been brought to England. This precaution would not prevent the Hawaiian monarchs Kamahameha and Kamamalu from dying of measles while on a visit to England in 1824.

The list of the epidemic outbreaks the inhabitants of Tahiti and the Society Islands suffered from the days of contact onward is a catalogue of the diseases that plagued Europe, but with even more horrific results in Tahiti, including influenza (1772-74), tuberculosis (1775), influenza, again (1820), whooping cough (1840). Between 1843 and 1850, eleven epidemics struck the island, notably scarlet fever in 1847, the same year nineteen ships visited Tahiti. In 1854, the measles killed seven percent of the population of the Society Islands (Kirk, 2011:290).

Finally came the universal scourge of the 1918 influenza epidemic, which was even more devastating in Tahiti than everywhere else. It was introduced in November 1918, when the *Navuna* arrived in Tahiti from San Francisco with several passengers ill with influenza. It is reported that one seventh of Tahitians died during the pandemic, around thirty times the ratio of deaths in France, with the older people dying in even greater number. Only the enforcement of severe travel restrictions prevented the epidemic to be spread to the Marquesas, the Tuamotus, and the Austral Islands.

Hawaii's Polynesian population was not spared either and suffered a similar decline. In 1778, the islanders numbered approximately 200,000; a century later, the numbers had fallen to 54,000. Syphilis and gonorrhea caused sterility and stillbirths, but the main reason for the decline was the advent, as in Tahiti, of devastating epidemics. Cholera (1804), influenza (1820), mumps (1839), measles and whooping cough (1848-9), and small pox (1853). The rapid growth in population starting in the 1870s was deceptive, as it was mainly due to the immigration of Asian workers hired by the sugar plantations (Kirk, 2011:386-7).

Some epidemics started for heartbreaking reasons. The islanders of Rapa-Iti, in the Austral archipelago, were the innocent victims in 1864 of a vicious action of Peruvian slavers, who were returning 470 indentured slaves from various mines and plantations to their Polynesian islands. The *Barbara Gomez* was sailing from Callao when smallpox was discovered among those on board. While on their way to Rapa-Iti, 439 ill or dead passengers were tossed overboard. Upon arriving, the crew told the Rapans that if they did not accept the handful of survivors, they too would be thrown into the sea.

The compassionate Rapans accepted the survivors—not all of them their compatriots—with disastrous consequences. Given their extremely low resistance to foreign diseases, three years later their numbers had dropped to an all-time low. When the first foreigners arrived in 1824, the population count was estimated at some two thousand. By 1836, their numbers had fallen to 453, mostly through disease brought in by visiting ships. In 1862, two years before the arrival of the *Barbara Gomez*, their numbers had fallen even lower to 360. When their numbers were counted again in 1867, the population only reached a grievous 120 (Kirk, 2011:343).

The epidemics in Polynesia had been caused by the islanders' lack of immunity and by unavailable or insufficient treatment. Carelessness on the part of the Europeans in not strictly enforcing quarantine among themselves when suffering from the disease—often benign for them, almost always fatal for the natives—is also likely to have been the cause of its spreading so far and wide.[15] It was an avoidable heartbreak, probably unintended, even if the Peruvian slavers were often guilty of cruelty in their handling of stricken Polynesians.

More recently, the French nuclear tests appear to have caused a number of serious health problems. Within a decade of the start of the tests (1966), a number of new conditions became apparent, particularly leukemia, brain tumors, and thyroid cancer. Once more, European activities were bearing their fruit, and the Tahitians' forceful opposition to the tests was rewarded too late. By the time France agreed to stop the experiment (1996), the harm had been done.

Finally, keeping up with the rest of the world in its menu of communicable diseases, the Tahitian health services have established prevention, diagnostic, and treatment centres for HIV-AIDS in Vaitavatava-Papeete, Touamotu Gambier, Fare Tama Hau, Tavaro, Moorea, Raiatea-Uturoa, and Taiohae-Nuku Hiva.

Population Count

There have always been fluctuations in the islanders' population counts. Initially at least, there were traditional reasons for the variation. The tribes were pugnacious and frequently at war with one another. In 1769, John Gore, an officer who had sailed with Wallis and was now with Cook, noted that the population had fallen within a year, and attributed this to wars being fought among tribes. He also noted other consequences of warfare, including burnt and demolished houses, and none of the pigs and chickens being in evidence although both had been found in abundance the year before (Kirk, 2011:59).

However, nothing would wreak as much destruction to the population as the diseases introduced by successive ships, causing numbers to decline dramatically. In addition, the demoralization of the people and the slave labour to which they were also subjected contributed to the loss of life. This combination of factors had created havoc among the people, whose numbers declined drastically.

In 1772, Johann Reinhold Forster had estimated the local population to be at least 200,000 people, although other figures only show about 50,000. Sixteen years later, Morisson guessed that there were 30,000 inhabitants left on the island. By 1797, Captain Wilson estimated that the population was only 16,000, suggesting an exponential decline within the previous nine years. These passing comments make it difficult to assess with any hope of validity the number of people living in these communities at the time of contact, as the numbers are often mere estimates and comparisons between periods are not accurate. However, what should be retained is the general, consistent, and ubiquitous downward trend.

Some of the information comes in the form of impressionistic comments. Thus, William Robertson (1954:24), master of the *Dolphin,* seeing the whole shoreline full of people, thought the island to be "the most populous country" he had ever seen. More reliable was Tupai, the high priest from Raraitea, who gave Joseph Banks a list of the districts in Tahiti with the number of warriors each could raise in times of conflict, allowing him to make a rough estimate of the island's population.

Other sources of population count came from the missionaries. F. A. Hanson estimated that about 50,000 Polynesians lived on Tahiti in 1767, when the Europeans first arrived. Thirty years later, English missionaries put the total at only 16,050. This number was further reduced to 8,568 (Tahiti's 1829 census). Other islands, likewise thickly populated when first seen by white men, underwent similar declines.

According to Hanson, in 1836 only 453 were still alive on Rapa-Iti, from some 2,000 eleven years earlier, fallen victim to the diseases brought in from visiting crews. "Nearly every visitor during these fatal years describes the Rapans as a wretched lot, diseased and dying in bewilderment and despair" (Hanson, 1970:27-30). By 1862, the population was down to 360, and when France annexed the Austral Islands in March 1867, only 120 Rapans remained .

As happens when using a variety of informal documents, numbers do not always match exactly. What is undeniable, however, is the regularity of the findings: that the population decreased in the most alarming way. Indeed, while it may be difficult to calculate this decline accurately by using different sources at various times, the figures obtained from the Gambier Islands (mainly Mangareva) with the help of official French records for 1840-80 are convincing enough: they show the loss of more than three quarters of the population, mainly through what can now be seen as slave labour.

The direst figures come from the Marquesas, where the original decline is mainly due to diseases. The population count in the late eighteenth century was estimated at around 60,000. In 1856, if had fallen to 12,500, then to 6,200 by 1872, further falling to 4,279 in 1897 (Dodge, 1976:198-99; Kirk, 2011:448). Like the population of Tahiti, it was to rise again slowly throughout the twentieth century.

The population count for Tahiti alone, starting with the 1767 rather fanciful estimates (from 50,000 to 200,000, from a variety of sources, including Cook's) is too vague to be taken seriously, although it does indicate a thriving population. Starting from more solid and more modern bases, such as the census, we can estimate the numbers to be 16,000 in 1797. By 1848, the population had fallen to an all-time low of 8,600, followed by a very slow increase until it reached 16,800 in 1931, the same level as in 1797.

If we are to accept the lower estimate of 50,000 inhabitants in 1767, thirty years later the population had already decreased by two thirds, a far cry from the decimation one assumes to be the worst ratio to be borne by a population. The year 1797 is coincidentally the one during which the London Missionary Society arrived in Tahiti, where the missionaries would have found an extremely vulnerable population already the prey of disease and other misfortunes.

Overall, the population's decline was so heartbreaking that Colin Newberry wrote about the 1911-1921 decade that one could be "pessimistic about the chances of 'native' inhabitants of holding their own in the long battle with infant mortality rates, infertility, and low resistance to imported diseases" (1980:271-72). During that decade alone, the total population of French Polynesia had fallen from 31,376 to 28,895. In the Marquesas, the numbers went down from 3,116 to 2,300. The Tuamotus had a similar drop, from 3,715 to 2,676. It is only after 1921 that the census numbers stopped dropping and one could anticipate them to rise across the islands.

Indeed, after 1931, the population had started to grow slowly but steadily. Between the counts of 1977 and 1983, Tahiti reached and passed the 100,000 mark. The 2007 census brings the population to slightly over 178,000. A study produced by *l'Institut national de la statistique et des études économiques* and *l'Institut de la statistique en Polynésie française* shows that in 2007 the figures for French Polynesia were 260,000, an approximate doubling of the population over the previous thirty years. It is believed that the increase was due to natural causes: more births than deaths. However, no such growth is expected in the future. As elsewhere in the world, the population is ageing, and those aged sixty years and over who constituted nine percent of the population in 2007 will very likely increase to seventeen percent by 2027.

Today, Tahiti's racially mixed population falls roughly within four ethnic groups: sixty-five percent are Ma'ohi, sixteen percent *demis*, five percent Chinese, and twelve percent Europeans. The last two groups live mostly in Tahiti and the urban areas. These more or less official figures may not entirely account for the various types of *métissage* that includes descendants from other races who had intermarried, or at least had conceived children with local women.

In the Hawaiian Islands, a population growth similar to the one in Tahiti did not reflect the actual Polynesian count. By 1876, the immigrant Asian population workers started boosting the numbers. The population growth continued, and in 1930 the US federal census showed a healthy total of 368,000, corresponding to an increase of forty-four percent within a decade. However, this total increase did not reveal the decrease (around fifteen percent) of the Hawaiian and part-Hawaiian population. By then, the Japanese constitute thirty-eight percent of the total population.

[VIII]
AND OTHER WOES

THE MISSIONARIES WERE NOT THE only Europeans whose presence was to affect deeply the life of Tahitians. It took very little time for traders to arrive, as the commercial exploitation of the Pacific began within decades of Cook's death in 1779. These were to be followed by whalers and settlers.

The Itinerants: Traders, Whalers, and Others

From the start, the Europeans and others who traded with China were eager to exchange the tea, silk, brocade, and porcelain they imported from Asia for what the Pacific islands produced in quantity, including pearls, tortoiseshell, sandalwood, and the delicacy of sea cucumbers prized by the Chinese. These transactions were often described as a booming "two-way trade" between Europe and Asia, and the irony seemed to be lost that there was a third party involved in the trade, the one who actually provided whatever the Chinese required and who were getting very little in return: the Polynesian Islands.

When the traders came in the early nineteenth century to gather sandalwood, the vegetation was stripped bare, the hillsides were denuded, and the soil was soon entirely eroded. Those who came for the guano, widely used as fertilizer, and mined the phosphate, often having to scrape off the surface and plants, made those islands inhabitable (Kirk, 2011: 684). The whole landscape of some islands was devastated.

The next to come were the British, American, and French whalers, who followed the whales from the Aleutians to the Antarctic for their oil and blubber. During the off-season, the whalers would sail to the small Pacific islands to refit, render and sell their oil, and relax. They also took on provisions of fruit, vegetables, and meat. While the navigators had traded nails, knives, and trinkets, the whalers brought muskets and rum. While the navigators had brought venereal diseases and tuberculosis, the whalers introduced other new diseases against which the natives had no protection either, including measles, cholera, and the common cold, which often reached epidemic proportions.

The whalers soon found their way to the Marquesas, attracted by the abundance of fresh provisions and the availability of women. Danielsson (1957:40) describes the distressing conditions faced by the natives whom the whalers often threatened with bombardment if they did not provide them with provisions. There was also the danger of being shanghaied when more crew were needed. And, to force them into obedience, few warnings could be as convincing as seeing some of them being thrown into the ocean. The men who returned, often bearers of new diseases, had also become well versed in the use of more sophisticated weapons, and tribal warfare then became more murderous than ever before.

Whalers and others sometimes interfered in feuds among natives. The worst intervention was that of the American Captain Porter of the *Essex* during the 1812-13 war between the United States and England. Feeling honour-bound to punish a tribe who had resisted him and who eventually had to pay an indemnity of four hundred pigs, Porter had the tribe's plantations devastated and its huts burned down (Danielsson, 1957:44).

On Rapa-Iti, the Austral island that so often comes to our attention for the treatment it received at the hand of the Europeans, the missionary John Davies gave a reason for the catastrophic decline in population he witnessed, from three thousand when he arrived in 1825 to six hundred only five years later. During that same period, "sandalwood traders, pearl fishers, whalers, and beachcombers taught Rapans how to distill liquor and many Rapans became alcoholics. Some of those visitors brought diseases. An undetermined number of Rapan men left with the visitors to work as seamen, divers, and whale ship hands. Some failed to return" (Kirk, 2011:217).

The whalers were active all over the Pacific. Whale hunting had been a very common practice in coastal communities, from the Nootka people in the Pacific Northwest to the Basques from San Sebastián and Saint-Jean-de Luz, who are said to have set up the first organized whaling fleet. It was expanded as an industry by the English and the Dutch in the seventeenth century, and was later dominated by the Americans. By the early 1800s, whaling ships from New England, mainly New Bedford and Nantucket, constituted the majority of the fleets. Of the more than seven hundred whaling ships operating in the world in 1840, four hundred had New Bedford as their home port. Kept at sea for long periods, the crews often needed to anchor in the Polynesian islands, and about two thirds of the Pacific whaling fleet ended up in the Hawaiian Islands. In 1846, Lahaina harboured 429 whalers, and Honolulu 167. Elsewhere, the sheltered bays of Nukuhiva and Hiva Oa (the scene of Melville's *Typee*) were also popular harbours (Lever, 1964:33-36).

The crews were the same, whether coming to Hawaii or to Tahiti: "derelicts, ex-convicts, Indians, freed blacks, Azores and Canary islanders, Polynesians, and Melanesians" (Day, 1968:131-32). We may assume that they were not overly scrupulous when dealing with the native populations on shore, where all they sought was relief from their harsh and tedious work.

Slavery

Slowly, the commercial value of the Pacific islands became evident to the major European powers of the time—Britain, France, and Germany. Whereas neither the traders nor the whalers had established permanent residence in the Pacific islands, this was not the case for the planters, whose arrival often coincided with the establishment of Christian missions.

Coconuts could be grown on volcanic islands, whose fertile soil provided coffee, cocoa, and most of all, sugar cane. But manpower was required and when local workers could not satisfy the ever-growing demand, men from other islands or even from Asia were forced to work on the plantations. They were often treated as slaves and were traded in a traffic controlled by people known as "blackbirders," who mostly came from Peru. In some cases, local chiefs and their families were held hostages until a sufficient local work force could be raised.

Kirk (2011:344) reports that in the early 1860s over three thousand islanders were taken to Peru. Of these, only about five percent (157) survived and were returned to the islands, although seldom to their own island, and many died soon afterwards. The Peruvian authorities appeared to have known and done nothing about the overcrowded conditions and the lack of food on the return journeys.

There were small victories, though. Aware that Peruvian blackbirders were abducting natives to work in guano mines and plantations, the French authorities succeeded in rescuing two hundred islanders from being carried off on a Peruvian vessel in 1863. The same year, natives of Rapa-Iti managed with the help of three Europeans to take by surprise a slavers' ship, the *Cora,* and bring her into Tahiti, where the French authorities arrested her officers and sentenced them to ten years in prison.

Officially, the British admiralty could do little to prevent these practices since many of the workers were often kidnapped from islands not under British jurisdiction. However, in June 1872, the British Parliament passed *An Act for the Prevention and Punishment with Criminal Outrages upon Natives of the Islands of the Pacific Ocean,* known as the *Pacific Islanders Protection Act,* at the request of the Australians, outraged that nothing seemed to be done about blackbirding ships (such as one caught in 1869 with a large number of African children off the coast of Zanzibar). The act was an attempt at protecting the islanders from blackbirding and indentured labour and from being cheated by recruiters or ship masters. It also authorized the British navy to intercept slave ships, whose masters were required to obtain a £500 bond against mistreatment and abduction of workers (Kirk, 2011:370). Finally, the Australian government succeeded in stopping the blackbirding practice in 1904.

Not all planters used the services of indentured or slave labour. Others had resorted to bringing Chinese workers legitimately. William Steward's Tahiti Cotton and Coffee Plantation Company, established in 1865, brought in thousands of workers. When the company went bankrupt, being unable to compete with the thriving cotton industry in the southern U.S. states after the Civil War, the Chinese were left to their own devices. They became market gardeners or small merchants, and today appear well integrated. However, they were treated differently by the French authorities, being only granted French citizenship in 1964, while Tahitians had received it in 1946.

When Danielsson visited Papeete in 1949, he found that most of the merchants there were Chinese. "No business is specialized; all of them sell anything that their own inclinations or the turn of affairs dictate, and it is not unusual to see such diverse goods as refrigerators, hats, cheese, razor blades gramophone discs, spirits, padlocks, and rubber teats all mixed up in the dirty disorderly Chinese shops"(1954: 5). Spitz's character Laura Lebrun found while visiting a Chinese shop a real jumble "with an endless variety of tinned food, flasks and bottles, most of them bearing a label in Chinese characters... Vast cabinets with shelves containing clothing, shoes, slippers. Hanging from the ceiling are scuba masks, flippers, torches, backpacks, watering cans, spear guns... In a glass-in cased, tooth brushes, combs, hair clips, tubes of toothpaste, deodorants, tiny bottles of eau de cologne" (2007:100).

As one drives in the countryside today, their enduring industry is evident. It is not unusual to see a Chinese name over a shop similar to what we would call a "minimart" in Canada, but very likely one offering a greater variety of goods. It is easy to see how these shops have become an intrinsic part of Tahitians' lives, and we see them figuring in modern novels as places where people shop for all their needs and where they also meet.

France had encouraged farmers to settle in Polynesia. During the 1889 Universal Exhibition in Paris, the Colonial Department had issued pamphlets informing the readers that "The French Colonization Society provides extensive free assistance for farmers willing to settle in the South Seas. Its limited resources do not at present allow it to distribute free land in the eastern Pacific possessions, as will be done in New Caledonia. But the kind support given to the Society by the authorities makes it possible at least to arrange free passage for serious-minded settlers who are firmly resolved on emigration to a French colony" (Danielsson, 1965:30).

Changing Technology

Contact with a more technically advanced culture usually brought a rapid change in the local use of tools, clothing preference, and local customs, a phenomenon observed wherever contact occurred. The tendency was for natives to prefer more efficient tools or weapons, and to dress in more practical clothes that also required less time-consuming manufacturing. They also allowed their practices to be gradually influenced and finally overtaken by those of the conquering or colonizing powers.[16] So it was with the Tahitians, whose initiation to the marvels of then modern technology was sudden and overwhelming, particularly if we remember that their technology in the eighteenth century was similar to that of the Stone Age. All Pacific islanders shared the same characteristics. By the sixteenth century, they had advanced to a Neolithic stage of development, which means that they ground, rather than chipped, stones; that they used seashells both as tools and ornaments; that they also used adzes to produce anything from weapons to tools. Some weapons were made of wood and stone, and they also used leaves and tree bark for clothing, decoration, vessels, blankets, and sails. The animals they had domesticated were only dogs, pigs, and chickens, and they did not have beasts of burden, nor did they use the wheel.

Their culture was almost entirely oral and the only records they left were on human skin in the form of elaborate and informative tattoos and on stones in the form of often mystical petroglyphs. These images have survived to a certain extent: the tattoos through the drawings and illustrations made by Europeans artists, although some of their meaning is unknown today; and the petroglyphs and *tiki* can still be found, sometimes near *maraes,* often partly erased by time and nature.

The Polynesians had put upon the sailors to provide them with nails. Barter of iron for sex had been established from the first contact with Wallis, to whom such a trade had seemed so grievous. It became somewhat more regulated with Cook who had established a system whereby any man missing his tools or his weapons would see their value taken out of his pay, but there is little doubt that the need for iron implements dominated the exchanges between Tahitians and navigators. The nails and tools made them more efficient manufacturers, and the weapons made them more successful warriors in intertribal fights.

With the availability of more technically advanced goods, it was natural to replace traditional stone adzes and bamboo knives with iron axes and any other modern implement that could be obtained through bartering. Such trends are difficult to reverse.

Some islanders, particularly those often engaged in warfare, soon learned what could be traded for food. When the whaler *Phoenix* reached Hivaoa in 1823, everyone already knew that the Marquesans were "complete heathens and desperate cannibals." The rate of exchange by then was one musket for each dozen hogs taken on board, and flint and cartridges for the vegetables and fruit. William Dalton, the ship surgeon, noted that this was by then general practice and, while the natives had previous been armed with "only clubs and spears, now they have muskets, which they procured from English shipping in exchange for refreshments, Sandal-wood and Cocoa-nut Oil" (Druett, 2001:172). Dalton also remarked that the Marquesans were "continually at war one tribe against another tho' living only a mile or two apart" and such was their thirst for efficient weaponry that they would not hesitate to capture European boats, kill their crew, and seize their arms if they could not trade them for food.

Societies evolve naturally, and the driving force behind any change is to seek an improvement over existing conditions. Even if the Europeans had not discovered Tahiti in the eighteenth century and immediately started trading for modern goods, today's Tahitians would not be identical to what they were then. The islanders would have been in contact with other civilizations, and even if they had not, there would always have been a man or a woman to discover a new way (easier, more efficient) to do things and to improve on them. The islanders' worldview and their belief system would also have evolved in the natural way as the human mind is able to create and imagine new metaphors, new myths, and new rationalizations. Each of these new steps would have been incorporated into the existing whole at its own pace.

Unfortunately for Tahiti, the assault came from too many fronts and at too fast a pace for this to be done painlessly. When the evolution occurs as a consequence of an abrupt contact with an infinitely more scientifically advanced culture, and when this contact is further compounded by the confusion and imposition of a new and alien religion, it is difficult not to feel regret that the Polynesian islanders were not given time to come progressively to terms with the inevitable changes that life brings, and compassion for the way those changes were so abruptly forced on them.

The Ignoble Savage

The Europeans had at first believed that there existed a natural inclination towards licentiousness on the part of Tahitian women—although we know that, while sexual education and performance were an intrinsic part of the culture, not all women had sex with the sailors. However, it became normal for those who did to be "rewarded" by the sailors with various iron implements. It took little time, seeing the value of the goods bartered for sex, for the commerce to become a crass trade off, with terms clearly stated and argued over. While in traditional Tahitian society sex between unmarried people had been normal, the white men's arrival had turned it into a true and sordid prostitution.

Every captain who revisited shores previously discovered would note the change in behaviour. From Tahiti to New Zealand to Hawaii, a new form of venality and prostitution had developed. By the time the *Pandora* arrived in Tahiti in 1791, the friendly if promiscuous approach that had greeted the earlier visitors seemed to have taken a more venal turn. "At Nomuka, where beautiful girls were brought on board and bartered for broad axes, the quarter-deck became the scene of the most indelicate familiarities," wrote George Hamilton (Salmond, 2011:303). Granted, the mission of the *Pandora* to hunt down the *Bounty*'s mutineers may not have been received with the good will that had greeted previous ships, but the general mood had also changed. Moreover, the chief of Nomuka believed his people had previously been insulted by both Cook and Bligh and he did not feel very hospitable towards the English. In fact, the islanders had become downright hostile, but this hostility was perhaps not the main reason of the change of attitude; simply, sex had become a commodity.

Later visitors to Tahiti often commented on the degradation of the people. In March 1834, Dr. Frederick Bennett, arriving in Papeete on board the whaler *Tuscan,* noted the results of the "debauched" lifestyle of the Tahitian people. "In the slovenly, haggard, and diseased inhabitants of the port, it was vain to attempt to recognize the prepossessing figure of the Tahitians as proclaimed by Cook." He blamed the lack of control of traders and the inefficiency of the government for this sad situation. "The abundance and indiscriminate sale of ardent spirits, as well as the laxity of laws which permitted the sensuality of a seaport to be carried to the boundless extent caused scenes of riot and debauchery to be nightly exhibited at

Papeete that would have disgraced the most profligate purlieus of London" (Kirk, 2011:224).

Added to the terrible deterioration of health and culture on the islands, there was a substantial change in Europe in the way the natives themselves were seen. "This idea of the 'noble savage' mutated in the nineteenth century into fear of the region and abhorrence of the people," writes Michael Moran on the drastic change of perception from earlier fashionable thought. It appeared to have done so for several reasons. "For Europeans, the Pacific Islands developed into competitive spheres of strategic and scientific interest as well as becoming a spiritual and physical testing ground for the various denominations of Christian missionaries" (2003:329-30).

Europeans, both at home and in Polynesia, saw ritual cannibalism and even the presence of disease they had themselves introduced as constant threats. They believed Cook's killing in Hawaii to have been all the more horrible to have been done with daggers made from iron spikes originally presented as gifts, with the natives repaying generosity with murderous hatred. This was a simplistic view that either ignored or dismissed as unimportant the misunderstanding that had caused the skirmish at Kealakekua Bay. Europeans almost seemed to believe that the natives did not "play fair" in rejecting their presence with such violence, and neither was it deemed acceptable that they would sometimes fight back when faced with the rape of their land and the destruction of their culture. What would have seemed a natural behaviour for them in the face of foreign domination and its destructive consequences was construed as ignorant and brutal hostility on the part of the islanders.

Most Europeans had not set eyes on Polynesian natives until the Paris Universal Exhibition of 1889. There, on the Esplanade des Invalides, they were finally displayed to satisfy everyone's curiosity. However, unlike the more important colonies housed in lavish pavilions, French Oceania occupied far more modest accommodations, denoting their lower status in the administrative hierarchy that governed their subordination.

Indigenous populations from other continents usually lose some of their characteristic charms when exhibited in the cold and often uncharitable light of colonial Europe. They need the natural context and mood of their own land to show themselves at their best. This is particularly true of those coming from islands of innocent beauty where life is so easy as to appear requiring no effort at all. As well, by 1889, they held little in common with the people Bougainville and Commerson had met and described.

Moreover, the image of a New Arcadia had suffered a reversal. There were now definite changes in the way Europeans perceived the Polynesians. While former writers had sang the pleasures of life in South Seas islands, new ones, such as Joseph Conrad, took to depicting the moral destruction life in the tropics brought on white men. Similarly, the Australian writer Louis Becke wrote stories of moral decay and perversity among South Sea traders. He spoke of "that strange, fatal glamour that forever enwraps the minds of those who wander in the islands of the sunlit sea…" This is eerily reminiscent of Bligh's reportedly saying in *Mutiny on the Bounty* that he did not care whether his men contaminated the natives or whether the natives contaminated them, as it seemed by then to have become of form of prophecy. No longer enjoying the Arcadian status of noble savages living in perfect harmony with nature, Polynesian life came to be characterized in literature by waning cultures, degenerate islands, and dissolute Europeans romantics fleeing the fetters of civilization (Moran, 2003:329-330).

Such was the fashion of the times. Today, the fashion is otherwise, as once more the vision of tropical islands has restored them to beauty through the aggressive promotion of the modern tourism industry.

PART THREE

FROM OTAHEITE TO THE NEW TAHITI

"I have always been of the opinion that we have no right to impose our ideals upon other nations, no matter how strange it may seem to us that they enjoy the kind of life they lead, how slow they may be at utilizing their resources of their countries, or how much opposed their ideals may be to ours." (Franz Boas, Letter to the *New York Times*, 1916.)

"This great universe... is the true looking-glass wherein we must looke, if we will know whether we be of good stamp or in the right byase... So many strange humours, sundrie sects, varying judgements, diverse opinions, different laws and fantastical customs teach us to judge rightly of ours, and instruct our judgement to acknowledge his imperfections and naturall weaknesses, which is no easy an apprenticeship." (Michel de Montaigne, *The Essays*, Book I, Chapter XXIV, 1603)

[IX]
RESISTANCE, NUCLEAR TESTING, AND SELF-DETERMINATION

NUCLEAR TESTING AND SELF-DETERMINATION ARE the main focuses of the colonial and post-colonial periods in Tahiti, but before the islanders were forced to accept the presence of the French, they actively resisted it. This chapter covers the history of the early resistance, the French government's decision to use French Polynesia as a nuclear testing ground, and the political turmoil centred on self-determination and Tahitian independence.

We are familiar with the early contact history of the islands through the navigators' reports and the missionaries' relations, but we have no contemporary Tahitian version of the events, except for those that have been transmitted orally. A modern Tahitian version of the contact is Louise Peltzer's *Lettre à Poutaveri* [the Tahitian rendition of Bougainville's name], half ethnography and half novel, that is respectful of the chronology and offers a valuable interpretation of the events. As a modern work, it does not escape being affected by what has since transpired and by what we now know of the devastating consequences of contact. Not only did they introduce the natives to iron, liquor, gunpowder, and the pox, but through the alliances forged with the chiefs, they affected the political balance among them.

Koerpelien, another contributor to the Tahitian myth through his romance with Tuimata, wrote about the Tahiti's history (2009:231-37). Once again, Europeans, albeit a sympathetic one in this case, would take the lead in attempting to depict and explain Tahiti to Tahitians. However, given his close connections with several Tahitians, he probably reflected some of their views.

After describing the arrival of the early Europeans and the state of contentment in which the Tahitians then lived, he retold the events that soon followed: the white men, for reasons "unworthy of civilized nations," deciding to take Tahiti under their "protection"; the natives, unable to understand why this was happening to them, trying to fight off the strangers who were attempting to destroy their culture; then, the cannons' appearance and the end of the rebellion.

"Whatever action taken by the natives that the white men did not understand, they called treason, and every native refusal to rally the white men's cause was called sabotage." Laws were enacted in a language Tahitians did not understand, and these were laws related to their traditional custom of sharing their food, to their houses, to their animals, and even to their national costume, the *pareu*. "What could spears and arrows do against guns and cannons? And what use was the relatively advanced culture of the Tahitians when their country fell into the hands of a George or a Carlos [King George III of England and King Charles III of Spain] who wished to add it to their conquests?"

Reviewing the next chapter of Tahitian history, Kroepelien commented on the reign of Tahiti's "self-proclaimed king," Pomare, sparing neither the venal king nor his minions. The king "only served the foreigners and, when he was not drunk, only worked at translating the Bible." And worse kept happening. "Although they were unable to understand why, Tahitians saw hostages being taken among the best men of the island, and with little apparent reason some were killed or tortured in the most dreadful way."

Before even tackling the delicate topic of the missionaries' promises for rewards in the hereafter, Kroepelien laid the blame where he saw fit. "The image of the generous white man is blotted by murders, violence, arson, and destruction throughout the Pacific, and one of the first places where people witnessed what a white man's nation calls Culture, Law, and Justice, was Tahiti." He could write this with a pure conscience, as a Norwegian who did not share the guilt borne by other Europeans in the Pacific.

Kroepelien quoted a Tahitian's question to a missionary: "You speak of the good things that you brought us, but where can we see them? You brought us syphilis and tuberculosis, and Tahitians are dying. All we ask is to be able to live our life in our own country, without all the gifts you brought us, and without participating in this new world."

There is nothing in what he wrote that should surprise us. More interestingly, in the face of the long recital of woes Kroepelien presented us, in which he did not appear to envisage much hope for the surviving Tahitians, he concluded by quoting a proverb: *Te pato nei te 'apu i tei 'ite i te ta'ata o tei i te tatari*—The seashell only sings for the one who can wait.

Kroepelien ended his Tahitian history in September 1842, soon after the annexation of the Marquesas, with Queen Pomare signing the papers that would make official the French presence in Tahiti. Such a pattern was repeated in many parts of the world. Ronald Wright (2004:200) quotes George Gilmer, governor of Georgia in the 1830s, on the way to rationalize such a process: "Treaties were expedients by which… savage people were induced… to yield up what civilized people had the right to possess."

Those were certainly the facts in Tahiti and her islands, but they do not reveal the political and religious background, nor the passionate reaction of the islanders.

Distress and Rebellion

Assaults and reprisals took place in several parts of Polynesia, the most famous being, of course, the events at Kealakekua Bay in 1779. Whether it would have been memorable for its importance had Cook not been killed there is debatable, as skirmishes occurred everywhere, particularly in the early days. However, what occurred when the French decided to occupy Tahiti and her islands had a different character.

Salmond notes that there was a growing resistance over time among the natives against the presence of those who had abused their values and carelessly brought diseases and afflictions upon them. Moran confirms the islanders' increasing sense of revolt. "The imposition of imperial Western values, the introduction of explosive weapons, European diseases and European commerce reaped a harvest of thorns as native people answered the rape of their land with violence" (2003:328, 330).

Spitz comments on "the myth of the 'soft colonization' that will have you believe that France was welcomed with open arms by Polynesian populations. France was never welcomed in the Marquesas. There were wars. At Tahiti, there were wars. In the history books, there are a few lines that refer to a few skirmishes. It's like the Algerian war: it was not a war!" (Pernoud, 2012).

The war that was not a war did not last very long, in spite of the spirit driving the warriors. The following is, roughly, what the records say happened before the resistance was crushed by overwhelming forces. Queen Pomare, as a close ally of the London Missionary Society, expelled two French Picpusian fathers (Laval and Caret) upon their arrival from Tahiti in 1836. The French immediately asked for compensation, even sending a first gunboat to Papeete in 1838 and another the following year. Pomare had no recourse but to meet their demands (financial compensation, an apology, and future welcome for French Catholic missionaries). Father Laval remained in the Mangareva mission (Gambier Islands) pursuing his megalomaniac construction plans.

The French then appointed as their consul a Belgian by the name of Moerenhout. In 1842, during an absence of both the Queen and the English consul George Pritchard, Moerenhout succeeded in having four chiefs sign a petition for more French "protection." The French admiral Dupetit-Thouars accepted their request and made the Queen agree to the establishment of a French protectorate. By 1844, Pomare had fled to Raiatea and Pritchard had been deported to England.

Pomare, who had lost much of her authority through this political move, sought the help of her "dear Friend and Sister Queen" Victoria. "Commiserate with me in my affliction, in my helplessness in which my nation is involved with France... Do not cast me away, my Friend; I run to you for Refuge... my only hope of being restored is you" (Kirk, 2011:270-72). The Foreign Office had no intention of antagonizing the French king Louis-Philippe, and Queen Pomare's plea was not answered.

It is not difficult to guess the turmoil of the period, with Tahitians facing a powerful and well-armed colonialist opponent. Melville described the events of the summer of 1842, when he arrived in the Marquesas, of which the French had taken possession a few weeks earlier. "They had disembarked... about five hundred troops... employed in constructing various works of defence, and otherwise providing against the attacks of the natives, who... might be expected to break out in open hostility. The islanders looked upon the people who made this cavalier appropriation of their shores with mingled feelings of fear and detestation. They cordially hated them; but the impulses of their resentment were neutralized by their dread of the floating batteries... with their fatal tubes ostentatiously pointed not at the fortifications and redoubts, but at a handful of bamboo sheds, sheltered in a grove of cocoa-nuts!" (1982a:26).

In 1844, Fanave, a chief fighting to preserve Tahiti's independence, led the resistance against the French. Soon afterwards, he and his troops were defeated at the Battle of Mahaena. Fifteen French soldiers died, against 102 Tahitians. In spite of the odds, Tahitian guerillas would continue to harass the French until 1846. Their greatest success was in repelling the French attempt to land in Huahine, with substantial losses for the French: eighteen soldiers were killed and forty-three were wounded. In the meantime, as we saw, Queen Pomare had continued opposing the French protectorate, and temporarily fled to Raitea.

The year 1846 was fatal for the rebellion. A Tahitian guerilla force was defeated at Fort Fautaua, and the French won the battle of Punaruu. Both victories overwhelmed the Tahitians, and there ended their last attempt at remaining independent, at least on the battlefield. Pomare returned from Raitea and accepted French protection. She was then restored to her throne, having lost many of her former powers. Thus ended the Tahitian war that had lasted almost three years against a much better equipped and trained military force. It had opposed Tahiti and her allies (Huahine, Bora Bora, Raiatea), and some European settlers, including the Protestant missionaries, against the French government and a few Tahitian chiefs.

However, this was not the last time the Society Islands refused to toe the line and accept the French rule. In 1888, the last islands to be annexed were Raiatea, Tahaa, and Huahine, before adding Bora Bora to complete the French Society Islands colony (later to be officially known as *la Polynésie française*, French Polynesia). The Raiateans fought the hardest and the longest again the French, finally losing their war at the battle of the Avera Valley in 1897, where their chief was taken prisoner. The forces against them were overwhelming, with two French warships, the *Duguay Trouin* and *Ti Aube*, arriving from New Caledonia with two hundred French soldiers. The *West Coast Times* of March 17, 1897, reported that some of the Raiatean rebels were deported to New Caledonia, while others were exiled in the Marquesas. The men who were not deported were used to form road gangs to improve the roads on Raiatea.

Such resistance as was offered in Tahiti and in Raiatea must have been very difficult to organize. Certainly not through any flaw in the islanders' fighting capabilities, as they were adept at guerilla warfare. Their bravery was well known and, indeed, during the two World Wars they showed themselves to be disciplined and expert soldiers. Their endurance during their long early migratory voyages also showed them to be courageous, organized, and tenacious.

However, their history and geography would not have predisposed them to form a systematically organized resistance uniting the various islands and their clans. During the days of contact, they had been able to fight off the incoming white men when it suited them to do so; they did it in small groups and on their own ground. When the time came to take up arms against the French, they had no experience with the type of forces that landed in Tahiti and Raitea, with far more military resources than the earlier navigators had possessed.

There were possibly several factors that limited their ability to organize into a proper opposition force, as had the Spanish guerillas against Napoleon, the French *Maquis* during the German Occupation of World War II, or Abdelkrim's War of the Rif Mountain in the 1920s. None of these possessed, relatively speaking, better weapons than had the Polynesians, yet they inflicted serious and prolonged damage to their enemies. What the guerillas, the *Maquis,* and the Rif Berbers possessed were both a common and unwavering goal to fight against the invaders and an excellent communication system.

In Tahiti, one of the main drawbacks was the lack of united commitment on the part of Polynesian chiefs or kings to reject the invaders, combined with the society's hierarchic elaboration that forced the subjects to follow their leaders and obey their directives. Thus, once the chiefs or the king found it to their advantage to back the missionaries, the islanders converted in large numbers. The same thing happened when many chiefs decided through self-interest to sign the protectorate papers; their subjects did not object. However, when Queen Pomare, Fanave, or the Raiatean chief refused to comply, they found enough followers to mount a serious resistance that would only be subdued by all the resources of the French military.

A second problem consisted in the very physical configuration of their archipelagos. They were scattered islands, often several sailing days apart, historically and traditionally engaged in hostilities. It would have required determination to bridge the geographical distances between the islands and the cultural ones that had separated them throughout their history. Resistance is usually of the moment, arising almost spontaneously, and capable of uniting former traditional enemies, as was the case with the Spanish guerillas, the French maquis, and the Rif insurgents, able to come together with a single goal that temporarily overlooked prior differences and even antagonisms.

However, coming together was not usually a forte for these island groups, even when not separated by long distances at sea. For instance, when American captain David Porter attempted to claim the Marquesas for his country in October 1813, he anchored in Nuku Hiva to reprovision his ships. The two neighbouring groups, the Ha'apa'a and the Taipi, living across the same valley but divided by precipitous ridges, behaved very differently. After a token resistance, the former agreed to provide food for the Americans, while the latter fought them off savagely. The Taipi acquired a reputation for fierceness that allowed Porter to exercise extreme violence in his reprisals against them (Porter, 1822: II-103).

Worst of all, the Polynesians were by then demoralized, and their population was decimated. By all accounts, their spirit felt defeated. They had faced unknown diseases, declining population, loss of culture and social structure, and often suffered from alcoholism. In these conditions, how could they find the energy and the conviction to rise against those who now seemed so securely established on their islands? Very likely, Melville's description of their "mingled feelings of fear and detestation" was an accurate one as they looked upon the invaders and understood their determination.

Colonial Days[17]

Looking back on the evolving pattern of colonization and unscrupulous treatment of native populations, Gilbert comments that "the cost of white domination was heavy. Epidemics, tribal wars, massacres, slave labor, alcoholism, and the confiscation of land from the island owners had depopulated islands and sown a harvest of bitterness and hatred" (1973:312).

The Earl of Morton, Diderot, Lapérouse, Cook, and other fair-thinking men would have been appalled to see their concerns and predictions so harshly realized, but might not have been surprised. Acknowledging the damage done was all later Europeans could do. They could not undo what had been foreseen: that disease and gunpowder, crucifix, and dagger were indeed the lot to be borne by a once thriving culture.

In his aptly titled *Tristes Tropiques,* Lévi-Strauss takes us later in the history of Tahiti, and adds a modern slant to his reflection. "Today when Polynesian islands submerged in concrete have been plane carriers heavily anchored in the south seas… how could we pretend that the escape of travel could succeed in doing anything other than force us to face the most unhappy forms of our historical past" (2008: 25). This is well put, indeed, the colonialists' *mea culpa,* but the statement seems to ignore the counterpart of this vision of Polynesia—almost as if the islands were nothing more nor less than the object of the white men's reflected guilt.

When Queen Pomare died in September 1877, a mere figurehead by then, Tahiti no longer seemed to have the will, nor perhaps even the taste, to think of independence. The Queen's ineffective son offered no opposition when Tahiti and her islands became a French colony on June 29, 1880, as the *Etablissements français de l'Océanie* (EFO).

For the native population, the following years were a period of apprenticeship as a colony, a sorrowful time, particularly given that their numbers had by then dropped to an appalling six thousand. In fact, the EFO may not have been seen as a particularly desirable posting for civil servants, as Kirk reports that as many as twenty-four governors served between 1882 and 1914.

Between 1881 and 1888, France annexed the Gambier Islands and the Tuamotus. The Queen of Uvea (Wallis Island) agreed to join the protectorate, as did the chiefs of Futuna and Alofi islands, their territories now jointly known as Wallis and Futuna. At the same time, the French abolished the hereditary position of *ariki* (chiefs).

In an attempt at economic development, the French exploited the potash deposits at Makatea. They also believed that the island of Rapa Iti could become an important stop on transoceanic passages, but the opening of the Panama Canal in 1914 put an end to these hopes. The island's climate was not conducive to the cultivation of coconut and not even copra ships would stop in Rapa Iti, which was replaced by Papeete as the obvious port of call on the way to New Zealand.

France continued to expand her overseas territories in Polynesia, and the last three Society Islands of Raiatea, Tahaa, and Huahine were annexed in March 1888, soon followed by Bora Bora. We saw how the Raiateans would continue for several years to fight against the French to maintain their independence.

Europe would be at war seventeen years later, and the first contingent of 165 Tahitians left in August 1914 to be trained in Noumea, New Caledonia. *The Battalion des Tirailleurs du Pacifique* was formed, with all eligible Frenchmen conscripted and all native islanders able to enlist as volunteers. Two contingents of the *Bataillon* sailed from Noumea in June and in September of 1916. By the end of the war, 1036 Europeans and 1134 Polynesians had served. The death toll was heavier for the native troops: 162 Europeans and 374 Polynesians, with the majority of the latter (207) dying of disease.

In September 1914, two German warships, the *Scharnhorst* and the *Gneisenau* bombarded Papeete, causing extensive material damage. Both at home and on the French front, native Tahitians participated in the war or suffered its consequences.

The Second World War also saw Tahiti responding with courage. After the defeat of the French army, the governor of Tahiti officially backed the Vichy regime of Marshall Pétain. When, on June 18, 1940 in London, General de Gaulle issued on the BBC a proclamation that France would continue to fight the war, both London and Vichy called on all French colonial governments to declare their allegiance to their side. Almost immediately, Tahiti chose the Free French side: an unofficial vote showed that de Gaulle received 5,564 votes and Pétain 18. On September 2, Tahiti declared itself loyal to de Gaulle, as did New Caledonia.

The most noticeable aspect of the decision reached in Tahiti was its near-unanimity.[18] However, the reality was less clear-cut than it appears on paper. There was still some strong resistance against joining the temporary Gaullist government, declared a traitor by Vichy. The new EFO governor had promptly put in jail some supporters of Pétainist Vichy who had mounted a coup on September 18. An observer commented that "the Papeete prison became so overcrowded with argumentative Frenchmen that it was the liveliest place in town" (Kirk, 2011:534). This is not surprising, given the usual political crankiness of the French and the deep confusion that reigned at the time, with their country torn apart and siding with opposing enemies. On April 21, 1941, the first eighty volunteers marched to the waterfront of Papeete to board the *Monovai* for Noumea. These were the first of three hundred volunteers from Tahiti to join de Gaulle's Free French Forces.[19] The contingent from Tahiti was to suffer ninety-six casualties in North Africa, Italy, and France before returning to Papeete on May 8, 1946.

Nuclear Testing[20]

It is difficult to write about the Tahitians' opposition to the French government's nuclear testing in their territory as a separate issue from their desire for political autonomy. They were the two major concerns that dominated politics after the 1960s; they were concurrent, intertwined, and feeding on each other. Yet, their roots were different, being emotional in the case of independence and visceral in the reaction against nuclear testing, even if it is at times difficult to differentiate between them. However, one issue was eventually resolved, while the other is still ongoing.

Following the departure of the French from Algeria in 1961, Charles de Gaulle announced that France would conduct nuclear testing (previously done in the Algerian desert) in French Polynesia and created in 1962 the *Centre d'expérimentation du Pacifique* (CEP). Not surprisingly, the Tahitians objected, even when the "President-General," as he was known in Tahiti, promised that the tests would only be done when the wind blew away from populated centres. He is said to have promised, "There will be no danger, all the necessary precautions will be taken." When the Tahitians still objected, de Gaulle was widely reported to have told them, "Go tell that to Messrs Kennedy and Kruschev. If they decide to give up their nuclear armaments I will do the same"(Tagupa, 1976:14; Kirk, 2011:584).

Between 1962 and 1966, the face of Tahiti changed and would never look the same again, nor would Papeete go back to its former small-town status. French military and technical personnel landed by the thousands on the recently built Faa'a airport and destabilized the island's infrastructure. Although some fifteen thousand Polynesians found employment, the mass arrival created housing shortage and inflation; traffic jams occurred in once-sleepy Papeete; and new taxes were imposed. Most of the food had to be imported and Tahiti's exports were totally eclipsed by its imports.

On July 2, 1966, the first atmospheric nuclear test was carried out at Moruroa Atoll, in the Tuamotus. In 1967, the French authorities built an airfield on Mangareva in the Gambier Islands, the closest access to the atoll used for the tests. During testing, the residents of Magareva's main town, Rikitea, were taken to large sheds, while water was sprayed over their corrugated-iron roofs to "decontaminate" them (Kirk, 2011:593). However, to give some perspective, schoolchildren in North America were still told to take cover under their desks in case of a nuclear attack.

Between the 1960s and the 1990s, the two interrelated questions of the status of the territory and nuclear testing continued to dominate Tahiti's political scene. Between the first (1966) and final tests (1996), 46 had been atmospheric, followed by 150 taking place underground. In Tahiti, Algeria, and elsewhere, nuclear testing has been the topic of persistent enquiry and strong condemnation,[21] as well as concern over potential long-lasting health consequences. In June 1995, President Jacques Chirac stated that France would resume testing. However, since the Moruroa Atoll had suffered some fractures and there was concern that radioactive material might leak, the decision was reached to perform instead underground testing at Fangataufa Atoll in the Tuamotus. This led to the most violent opposition against the French since their annexation of Tahiti. It was reported than five hundred people occupied the runway at Faa'a Airport, spreading rocks and coconuts on the ground to prevent the planes' landing and taking off. They demanded a referendum on nuclear testing and independence for the island. When the police intervened, people smashed the airport windows, broke video monitors, and set fire to the restaurant and to cars in the parking lot. As night fell, the rioting reached the centre of Papeete and violence spread there (Kirk, 2011:650).

The day chosen for testing was particularly irksome. "They did not want to have it over French Christmas," said acting Australian Foreign Minister Gordon Bilney. "But they were quite prepared to make this a New Year's present to the South Pacific." In New Zealand, frustration also ran high.

"It's really quite incredible that France didn't listen to not only the South Pacific, but to the Commonwealth leaders and to the United Nations," said New Zealand Prime Minister Jim Bolger. "The voice of the world says no to nuclear testing, and you are left wondering what part of 'no' the French government does not understand" (CNN World News, Dec. 28, 1995).

In Tahiti, anger was at its peak. Any country forced to subject to nuclear testing against its wishes would understandably be up in arms. This is all the more true of Tahiti because of the visceral bond the inhabitants have with the land, which they saw being deeply harmed by nuclear tests through decisions made by people who had no understanding of their beliefs and who safely lived thousands of mile away.

When the decision came to stop above ground testing in favour of carrying them out underground, these actions took on an even more metaphorical meaning. The Tahitians' profound bond with their land is best illustrated by the custom, still practiced, of burying the placenta and umbilical cord soon after a child's birth. This symbolizes and confirms its attachment to the land.

Many writers have commented on this bond. Miriam Kahn develops the metaphor in a chapter titled "Placentas in the Land; Bombs in the Bedrock." Bruno Saura, another specialist of Polynesian civilization, has studied this ancestral custom in "Le placenta en Polynésie française" and asserts that the custom has survived in spite of hospital births and delocalization.

The idea is simple: nurture the land that nurtures you. The sea provides the fish, but the land sustains with food, shelter, and comfort. Burying the placenta forges the strongest possible bond between a child and the land that will nurture him, as his mother did through the placenta; wherever he may go, an essential part of him still belongs in the land of his birth. It then becomes clear that anything that harms the earth is a matter of utmost importance to the Tahitians. This is simplistically put, but the importance of the custom and the belief attached to it is not to be treated lightly and the language itself reflects the connection: the land, the earth, is called *fenua* and the placenta nourishing the baby in the womb is called *te pu-fenua*, the centre of the earth. Burying the placenta and planting a tree on it restores the cycle of life.

Spitz, writing about the departure of a young soldier to France, has his mother saying, "Tematua my son, on the day of your birth, I entrusted your *pito* (umbilical cord) to your land." Upon his return, Tematua will find this land to which he belongs because a part of him has been entrusted there. Thus attacking it with a deadly weapon, as the French were seen to do while testing underground, was an unforgivable action.

Finally, in February 1996, France acceded to the Test Ban Treaty and ceased nuclear testing. The situation in Tahiti had become serious, with riots in Papeete and severe boycott of French products. The French decision was greeted as a major victory by the Pacific nations opposed to nuclear testing. Concluding the chapter on French nuclear tests, even if the chapter is not closed for those still suffering from their aftermath, the law of January 5, 2010, called *loi Morin*, recognized the victims of nuclear tests in French Polynesia and Algerian Sahara.

Self-Determination and the *Indépendantiste* Movement.

After the war, the relations between Tahiti and France evolved rapidly. On October 25, 1946, the newly constituted French provisional government granted the islands representation to the French Assembly. Under the Fourth French Republic, all adult islanders were able to elect ten representatives from Tahiti and its dependencies, five from the Leeward Islands, and two each from the Tuamotus, the Marquesas, and the Gambier Islands. Two days later, the EFO and New Caledonia officially became the *Territoires français d'Outremer* (TFO), French overseas territories, each to be represented in Paris by their own *député* (elected representative) and their own senator.

Before reviewing the events that followed, we should note some significant landmarks along the way to the twenty-first century. In 1946, Tahitians received the right to vote; in 1957, the first local government was elected; in 1977, the Tahitian language, soon to be declared the second official language of the island, started being taught in schools.[22] However, the road to these successes was not easy, and French Tahitian politics seem as confusing and acrimonious to the outsider as metropolitan French politics are themselves. Moreover, they were always dominated by strong characters. One of them, and among the first to preach autonomy, was Pouvanaa Oopa (1895-1977), a *demi* born on Huahine, who had fought as a volunteer during WWI. During 1945-47, the anti-colonialist movement started organizing and in 1949, Oopa was elected and twice re-elected *député*. Disappointed with the little autonomy Tahiti had been able to obtain from the French, he founded his own autonomist party, the *Rassemblement des populations tahitiennes (RDPT)*.

The new party's platform sought to provide a liberal constitution for Tahiti; to replace French civil servants with local people, with the exception of the governor and a few technical agencies; to return to public custody a number of lands previously ceded to France under the colonial regime; to establish agricultural and commercial cooperatives; to created good working conditions and ensure a minimum wage scale; and to relax the constricting economic ties with France.

The opposing parties, lead by Charles Vernier and Emile Vernaudon, well connected to the French protestant church, had little of any substance to propose, and the results of the votes during the 1949 election campaign for the French National Assembly showed that about forty-five percent of the voters had endorsed the *RDPT* platform. Opposite him, as well, was Rudy Bambridge, founder of the *Union tahitienne démocratique (UTD)*, a party supporting French sovereignty.

Pouvanaa Oopa became the first vice-president of his party and head of the local government of *Polynésie française,* so renamed on July 22, 1957, and including the five archipelagos of the Society, Tuamotus, Marquesas, Gambier, and Austral Islands.

The physical resistance that the individual chiefs had not been able to organize in a previous century had found in the twentieth its voice and some of its modern heroes. However, there was little unanimity in Tahiti on the best course to follow. On September 28, 1958, de Gaulle scheduled a referendum in French Polynesia giving voters several options: (1) to maintain the status quo, (2) to become a *département* of France, (3) to become an autonomous member of the French community, or (4) to become an independent state by rejecting the French Constitution.

Bambridge's *UTD* wholeheartedly endorsed "the maintenance of the French presence in French Polynesia under the guidance of French authority and not under some sort of republic." Oopa's *RDPT* was less convinced that it was the right thing to do, but the impending formation of the French Fifth Republic had led some to believe that this new government might be more liberal in its handling of overseas territories.

De Gaulle himself had defined the issues:

> "If you say YES to the referendum, France will consider you have accepted to remain with her for better or for worse. You have determined yourselves the manner of your internal independence in the fraternal French community. I you vote NO, France will know that you have chosen to leave the nest and she does not expect you to return. She will wish you luck and cease all material aid since you have you have considered yourselves capable... of earning your own way by yourselves" (Danielsson, 1974:25-26; Tagupa, 1976:9-10).

In the face of such an ultimatum, Pouvanaa Oopa and Jean B. H. Céran-Jérusalémy, an ambitious *demi* who had redacted the political newsletter of the party, reacted differently. Pouvanaa, reflecting the wishes of rural areas, strongly favoured independence, away from the *popa'a farani* (the French administration); Céran-Jérusalémy, supporting the urban faction, endorsed the continuation of economic ties with metropolitan France, believing it would be essential to the future economic prosperity in the island. The rift between the two increased to the point of leading Céran-Jérusalémy to creating a splinter party in May 1958, the *RDPT-Aratai*. On the other hand, *RDPT-Comité-Pouvanaa* remained unchanged in its politics and, in response to de Gaulle's ultimatum, issued the following statement:

> "If we answer YES we will continue to remain under the colonial enslaving government which we have known for seventy years... Vote NO so that the enslaving yoke will be quickly removed from around our necks... The people should show gratitude to General de Gaulle's statements... where he said the time would come when we will be able to govern ourselves for our own good. Vote NO because we have been given true freedom by General de Gaulle's statement to vote NO. Vote NO to ease France's burden in French Polynesia. The *RDPT* chooses NO" (Tagupa, 1976:10).

In spite of this urging by the *RDPT,* a majority of sixty-four percent voted to remain part of French community. The Fifth French Republic had just been constituted, resulting in an increase of the governor's powers to the detriment of the *Polynésie française*'s local government. As a result of the referendum, Pouvanaa Oopa's movement suffered a major setback. He was considered a threat and, despite having immunity as an elected representative, he was arrested following some civil unrest caused by his supporters and sentenced to eight years in jail and fifteen years in exile.

In spite of his absence, the *RDTP* continued to be supported by the voters. Pouvanaa's son had replaced him as the head of the RDTP, in spite of having health problems (he would die in 1961), while Céran-Jérusalémy continued with the *RDTP-Aratai*. However, political rivalry among the various parties continued through the 1962 election and cut through the support for the *RDTP*. The latter won the election, but with a smaller margin than before.

France was once more reviewing the status of her Polynesian territories and between 1946 and 2004, the archipelagos officially became *Pays d'Outremer au sein de la République (POM)*. They were now deemed to form an integral administrative part of France overseas. Two other territories rallied with a similar renaming: Mayotte, in the Indian Ocean, and St Pierre and Miquelon, off the east coast of Canada near Newfoundland.

In Tahiti, new important developments occurred, focusing on political autonomy, as ever the topic of passionate interest and a source of continuous unrest. New parties were formed. *L'Union tahitienne* (founded by Rudy Bambridge, later led by Gaston Flosse) rallied the Gaullist party, while the more autonomist and radical *Rassemblement des populations tahitiennes* (RDPT) was finally dissolved in 1963. It was replaced by the new *Pupu Here Aia* party in 1965, and led by John Teariki. The same year a new political personality appeared: Francis Sandford, the leader of a new autonomist party, the *Te Ea Api*.

In 1968, Pouvanaa Oopa came back from France and was received in triumph in Papeete. In spite of being in poor health after the years spent in exile, he became senator in 1971, representing the *Pupu Here Aia* party. During the next few years, most changes were aimed at promoting greater internal governmental independence, and in 1984, France agreed to increase the territory's autonomy, while still reserving control for foreign affairs, finance, defense, and justice. Universal suffrage guaranteed the vote for all French residents in Polynesia and all native Polynesians.

Writing about the 1970s, Donald Topping believes that, as far as other Pacific Islanders were concerned, the French Polynesians who voted for the status quo in their relationships with the Métropole had been "bought out." He argued that jobs with good salaries were more important to them than the satisfaction of being independent, and gave as an example of this search for good incomes the movement towards urban centres. Most went to Papeete and suburbs, whose population grew by ten percent between 1967 and 1971. Nearby Faa'a and Pirae grew by nearly 185 percent and 126 percent respectively between 1962 and 1971, the rate still increasing as he was writing (1976). By then, more than half of the French Polynesians were living in urban areas, where the jobs were. Topping cites other population movements (to the nickel and building industries in New Caledonia, where high salaries could be earned) and their consequences (the rise of *bidonvilles* on the outskirts of the cities), and concludes that they served "to accelerate the change from a Pacific-style rural subsistence economy to a cash-based economy… Traditional values and identification are badly strained as Polynesians leave their home islands and families to follow the job market" (1977:25).

Within four years, two symbolic events occurred. First, the French Polynesian flag was adopted in 1984, consisting of three horizontal stripes: two red outer stripes and a wider white central one. The territorial symbol, a red and brown Polynesian outrigger canoe, is located in the middle of the flag. Five sun rays are on each side of the canoe's sail, five men are on board the canoe, and five layers of blue sea support the canoe—each series of five representing the territory's five archipelagos. Second, a patriotic song, "Long Live Tahiti Nui," the closest thing that Tahiti has to a national anthem, was officially adopted by the Territorial Assembly in June 1993. Its first public performance occurred on June 29, 1993, during the celebration of the ninth anniversary of the internal autonomy statute of 1984. The lyrics translate into "God created my country/garland of multiple islands/ with such delicate fragrances/linked up as an everlasting braid/today let me praise you./Listen to your children's voice/crying out "Lavish your Love/ So that Tahiti Nui can live."

During 1990s and 2000s, politics focused on problems of home governance. The population was changing and had become increasingly urban and working class, and this cultural and economic development was influencing local politics. Tahitian political life was structured around two parties. One was the *Tavini Huiraatira* (People's servant party), an autonomist group founded and led by Oscar Temaru, a long-time mayor of Faa'a, Tahiti, and a former strong opponent of nuclear testing. He has served as president of French Polynesia off and on since 2004, and sought an increase in the minimum wage, an improvement in social services, and education, while still aiming at increasing the local government's autonomy. The other party, *Tahoera'a Huiraatira,* was led by Gaston Flosse, Oscar Temaru's long-time rival.

In 2004, several groups gathered around *Tavini* and formed the *Union pour la démocratie* party *(UPLD).* Between 2004 and 2014, the political arena seemed to suffer from chronic instability, as twelve governments followed each other, with the same three men alternatively presiding: Gaston Flosse (four times), Oscar Temaru (five times), and Gaston Ton Sang (three times). The current (2015) president is Edouard Fritch, of the *Tahoera'a Huiraatira* party.

To an outsider, modern Tahitian politics seem to have somewhat moved away from their early noble independence-driven aspirations, and the time seemed far away when Oscar Temaru was proud to call Pouvanaa Oopa *Oouvanaa Te Metua,* "Father of the nation," and describe him as "the father of Tahitian nationalism." The new politics appear to be frustratingly moored in petty competition and self-interest. It is disappointing to read the two Tahitian newspapers, *La Dépêche de Tahiti* and *Les Nouvelles de Tahiti* and see the niggling attacks on political opponents. Perhaps outsiders should not attempt to penetrate the arcane vagaries of internal politics or what may appear to be merely nuances rather than the profound differences they often are, or vice versa. Politics are passionate affairs best understood by their practitioners, their analysts, and those who live with the consequences of the politicians' decisions.

As we consider the strong political impulse towards independence felt by some, we see that it is paralleled by a literary movement starting in the 1960s and 1970s that found its inspiration in the epic poetic past of the island.

[X]
LANGUAGE, LITERATURE, AND POLITICS

Bilingualism and Translation

I have lived with Canada's official policy of bilingualism and biculturalism for nearly six decades. I reside in British Columbia, where the second unofficial language is Cantonese, and I occasionally visit Quebec, where the everyday language is French. I spent my childhood and adolescence in Morocco where Berber, Chleuh (a Berber dialect), Arabic, Yiddish, French and, since we lived near Ifni, Spanish were commonly spoken, but not necessarily by the same people. Finally, I went to school for a while in the north of France, where people spoke Ch'ti (also known as chtimi), a dialect despised by the French until a successful film made it quaintly acceptable in 2008. Even if I tried, I could not ignore the politics of language and literature.

Before considering the Ma'ohi/French situation (opposition, for some) in Tahiti, I suggest we review a fifty-year-old study on the co-existence of two languages and cultures in Paraguay, Guaraní and Spanish, because the circumstances were quite similar to those in Tahiti and may give us some understanding of the staying power of some native languages.

Unlike the multiplicity of native languages of Central and South American that eventually lost their currency, Guaraní (spoken by a homogeneous indigenous population in Paraguay) was such a resilient language that it was able to compete with the colonists' Spanish. There are many similarities with Tahiti where a well established language, similar to those spoken on the other islands, survived to the point of still being spoken by everyone in spite of the considerable inroads made by French, the language of officialdom, administration, business, scholarship, tourism, etc. Most importantly, French is also the language of literature, where the spirit of a people expresses itself if it wants to be understood beyond its borders. The usual oppositions between the use of Guaraní and Spanish (rural/urban, familiar/formal, older/younger, lower/upper class) and the preference for one language over the other in these categories is the same as the Tahitian and French linguistic duality.

The point about Guaraní-speaking people is that, alone among other South Americans, they have retained their language. The linguist Rubbins Burling makes the interesting observation that "some countrymen switch to Spanish when they are drunk since Spanish is the language of power, and alcohol confers a sense of power." Conversely, "even upper-class Paraguayans generally like to speak Guaraní if they go abroad, for then Guaraní becomes a sign of distinction, a point of national pride" (1970:101). In these two very specific examples, it is once more confirmed that power and national pride are symbolically attached to the choice of a particular language.

It is legitimate to wonder how the Guaraní language is faring in the twenty-first century, since the study is, after all, fifty years old. Today, Guaraní is enshrined in the Constitution of Paraguay; it was recognized in 1992 as the country's other official language, giving it equal footing with Spanish, the language of European conquest. It is also a source of national pride on the street.

The reasons for the survival and very good standing of the language may be due to several factors. First, because of its geography the region generally escaped the effects of the *encomienda,* the Spanish system that forced natives to work for Europeans, a situation that would have negatively affected the native language.

Second, the Jesuits created communities for the Guaraní and armed them against slaving expeditions, while also using their language in books and sermons. After the Jesuits were expelled by Spain in 1767, Guaraní spread throughout the country and, later, its speakers supported the post-independence ruler José Gaspar Rodríguez de Francia. Later dictators would also seek the backing of the Guaraní, who became *ipso facto* a political force in the country.

Another factor for the language remaining so successful was the region's isolation. The Paraguayan novelist Augusto Roa Bastos, who mixed Guaraní with Spanish in his writing, called his landlocked nation an "island surrounded by land."

There are other similarities between the Guaraní and Ma'ohi linguistic evolution. Antonio Ruiz de Montoya and other missionary-linguists transcribed the language by using the Latin alphabet, creating a written form, and some new terms had to be invented or transformed to conform with the Christian reality of the time. Thus, *Tupn* (Great Spirit) and *Karai* (sacred) were extended to refer to God and Christian. Over time, other transformations occurred, often in the pronunciation. The language "still had the same roots and speech patterns as Guaraní, but many words had been borrowed from Spanish," writes Nathan Page. We see the same borrowing from the dominant colonial language happening in Tahiti.

Guaraní writers who wish to reach an audience outside their own country must write in Spanish, but they still include enough Guaraní words to create a sense of belonging to their original culture—as do Ma'ohi authors writing in French but using native words throughout their texts.

Interestingly, and confirming its being accepted as a valid and current language with literary possibilities, foreign works have been translated into Guaraní. One would expect the *Book of Mormons* to have been, but more convincingly, so was *Don Quixote*.

In spite of its relative success, not everyone is certain that the future of the Guaraní language is safeguarded by these measures. However, for the time being, it is extensively used on the street and in that universal language of politics, the graffiti. Today, whatever the future may hold, to be accepted in Paraguay means being able to speak Guaraní.

In Tahiti, native writers must use French in order to reach a wider audience, and, perhaps even more than the Guaranís, some bitterly resent being forced to resort to "colonized writing." Spitz, in particular, is vehement about the constraint of *"écrire colonisé"* and wishes to escape from the "trap of francophony." She blames history for the accident of her birth in twentieth-century Tahiti, and claims feeling no kinship with the French just because she has to use their language (2002:2). She dreams of a new "navigation" across Oceania, as in the olden times, when canoes crossed the seas to create new settlements. This modern and literary navigation would connect islands that now have a different colonial language imposed upon them, be it French or English. She dreams of a pan-Oceanic intellectual movement, with purely indigenous terms of reference, and using the Tahitian language, *reo ma'ohi*.

Reo Ma'ohi

Originally used to designate native plants and animals, the term *ma'ohi* has since taken on a political connotation. Miriam Kahn (2011:9) traces its evolution from the late 1940s, when it was first used by the supporters of Pouvanaa Oopa to promote nationalist sentiments. In the 1960s, it acquired a new flavour when it was adopted to oppose the hotly debated topic of the French nuclear testing program. With the cultural revivalist mood of the 1970s, it acquired yet another strong impulse, particularly with Henri Hito (1944-1990), the voice of the Ma'ohi independence movement. It continued to maintain its political identity in the 1980s, when the expression *"Te fenua ma'ohi"* was used to denote the homeland and reject even more strongly French colonialism.

Nowadays, *reo ma'ohi* is spoken by all islanders and is taught in school as the other national language. The weekend edition of the newspaper, otherwise written in French, devotes one page to articles, poetry, and crossword puzzles in *reo ma'ohi*.

To linguists, Tahitian is an East Polynesian language, a part of the large Austronesian linguistic family, and closely related to its neighbours, Marquesan, Pa'umotu (Tuamotu), Austral, Mangarevan (Gambier), Maori, and Hawaiian. Actually, it is related to languages spoken in even more distant lands and appears to have spread across the Pacific Ocean during long migrations, since Austronesian languages were spoken wherever ocean-going canoes were found. Joseph Banks, a philologist as well as a botanist, had carefully looked at compilations of Polynesian words and believed that Tahitian was directly related to other languages spread across the Pacific to the Southeast Asian islands of the "East Indies." This is why some writers draw their political and artistic momentum from this close linguistic connection among Polynesian islands where the language of published literature is European.

The language spoken in Tahiti, admired by Commeson as "very sonorous, very harmonious… without syntax, suitable for expressing all their ideas and all their needs,"[23] was first written down by the missionaries, who simply matched the sounds they heard to the letters they knew. To the five vowels, they added the only consonants that seemed to apply phonetically: f, g, h, m, n, p, r, t. v, with the addition of an apostrophe denoting a glottal stop *(eta)* at the beginning of the syllable that follows it. It is considered to have the value of a full consonant.

The language has undergone many changes since the time the Europeans arrived in Tahiti. By 1918, Kroepelien (2009:178-180) already thought that "the language spoken in Tahiti today is so different from the one practiced in the olden days that a Tahitian could not understand the speech of his ancestors. Modern Tahitian is concocted in Papeete, where the contact with foreigners has been the crudest." For this, he partly blamed the missionaries who did the transcriptions. He also commented on a common expression, *'aita,* which means "no" and "nothing" and was probably introduced by the Chinese. According to one of his informants, the language spoken along the coast is *aita-maita'i,* meaning the "no-good" dialect and indicating that many foreign words had contributed to this situation.

Kroepelien was very much taken with the complexities and subtleties of the language, such as the ways of expressing the various components of the pronoun "we." When it means you and I, Tahitians use *taua;* when it refers to others and I, it is *matou;* finally, when he and I are included in the pronoun, but not you, the word used is *maua.* One is reminded of the hundred words the Inuit supposedly used to describe the condition of the snow. It simply means that what has important cultural relevance is named with great care and precision. The social structure of the Tahitians and their interpersonal relations obviously required the same care and precision in the naming of human exchanges, and whom they included and excluded in these transactions.

Kroepelien concludes, "Today, few people know old Tahitian, and it will not take long before these people disappear and, with them, the language. It is as if the Tahiti of the olden times was meant to die with the history of the Maoris." Certainly, much has happened during the century since these words were written and most linguists agree that today's language has been affected by the missionaries' transcriptions; by the introduction of new events, artifacts, technologies, and situations that required the creation of new, often metaphorical, descriptors; by the natural erosion of time; and by the contributions of non-Tahitians. All languages evolve, and those who run parallel to another and politically dominant one in everyday life are bound to evolve even more.

Today, Tahitians may not speak exactly the language known to Koepelien when they speak *kaina,* a franco-tahitian dialect which, according to Kareva Mateata-Allain, "modifies French prescriptions of syntax, pronunciation, and grammar." She sees in its effort to give the language a native touch a form of resistance to "the coloniser's language." Sometimes also called *franitien,* it is, she explains, "a colloquial French pronounced very much like Tahitian, with the trill of the rolled 'r' as in Spanish along with guttural vowels sounds" (2009:3).

The Oral Tradition[24]

The power of Tahitian oratory was recognized early by the Europeans. "These gentlemen, like Homer of old, must be poets as well as musicians," wrote Banks. Mateata-Allain confirms that "speaking and oration were highly privileged in pre-colonial Tahitian society, and this privilege is still entrenched in ancestral memory." Kroepelien also reminds us of the existence of the *haere po* (also the name of his publisher in Tahiti), whose function was to transmit the ancient tales throughout the generations, possibly going back several centuries.

There is a remnant of this old oral tradition in the form of the *'orero,* still practiced nowadays. The term refers to the traditional oratory form, and is familiar to old people who remember the stories of their ancestors. It has several meanings. It refers to the orator himself, the one who in former times assumed the function of a messenger for the family, the population, the king, and the gods. Throughout his childhood, he followed the teachings of *tahu'a* (priests) well known for their mastery of the culture. Once he had himself mastered this field, he was chosen by the *arii* (chief) as messenger, and recited appropriate texts during various ceremonies at the *marae.* It also refers to the manner of speech designed to captivate the attention and convince the audience. By extension, it also means the speech itself, the very discourse pronounced.

The oral tradition is usually full of rigour, and Tahitians adhered rigidly to its rules. Mateata-Allain reminds us that in pre-colonial society, Tahitians selected for the strict discipline of oratory were bound to excellence in their performance. No errors in syntax, structure, or pronunciation were permitted, for "a chief only had power if he or she maintained the gift of flawless orality in a society that privileged articulate discourse." To illustrate the difficulty of the task and the feats of memorization to be achieved, she explains that the Great Priest, *tahu'a nui*, supervised the recitations of several centuries of genealogy. "If the orator made a mistake, one lapse in memory, it would bring bad luck and devastation, and the speakers would develop a reputation as clumsy and inarticulate" (2009:6,9). A Spitz character remembers, "those gifted orators who spoke to them for hours without their concentration ever wavering... These men who had mastered their language so lovingly, filling their souls... These great speakers who have passed on to them the dreams of his people, showing them the colours of the world, the beauty of life, the love of their Land" (2007:35).

Today's *'orero* has lost its sacred and inspirational character. It is often used to accompany social events, such as weddings, where the genealogies of the two spouses have to be recited. Perhaps less arduous than in the past, the recitation must still remain a difficult one, with the pride of the two families and their respective ancestries at stake.

The transition from the oral to the written forms of transmitting texts has created for some writers a deep chasm in the way they process and retain information when using both forms. The poet Flora Devantine expresses this duality when she writes that ideas "seem to erase themselves/ almost simultaneously/in my mind./It is as if my brain/emptied itself out/at the same time as/the page filled up." In the second case, when the mind has to retain the words and express in such a way that it guarantees their memorization, "Like the craftsman,/.../ I turned and turned them over/giving them shape little by little/polishing them/as well as I could" (1998:17). Yet, she also makes the point that the written form serves to preserve the oral one. "We have forgotten a little/the particular words/ religious, sacred/of the language! (...) So we must resort to writing them/ which allows them to get their breath back" (1998:16-17).

Another of Devantine's arguments is that setting texts to paper could lead to their being forgotten in their oral version. Since this version is at the source of all Tahitian traditions and knowledge, their loss could be far superior to what is gained by their being transcribed.

I grew up in southern Morocco. At the market place, I often saw groups of people gathered around an old man reciting stories of long ago. The first rows squatted on the ground, the others stood behind them, and their torsos would sometimes sway slightly to the rhythm of the old man's speech, in which they had lost themselves. Those stories were acted out with intonations, modulations, gestures, pauses, or silences to highlight special moments. The text had the "breath" Devantine writes about, and even as a child I could see that it was alive.

I had first thought that the situation I describe in Morocco in the 1940s would not be comparable to the one in Tahiti nowadays, for those Berber participants-auditors could not write, and their only access to the epics they already knew almost by heart was through their oral rendition. I had assumed that the school system in Tahiti would have created a new literate generation. This is perhaps true now of the younger people, but Mateata-Allain wrote in 2003 that in Tahiti "illiteracy [was] pervasive."

Tahitian writers are only too aware that the passage from ancient form to modern expression creates a fraction in the literary impulse. All the more so that literacy in Tahiti came at the price of conversion. When the missionaries invented writing in Tahiti, it was not intended to transmit the old native tales. Yet, Louise Peltzer, a Tahitian linguist, is at some pains to establish that the members of the London Missionary Society took care to separate literacy and evangelization. She writes that there were in fact two schools, one "to learn to read and write and another school to learn the new religion" (1993:287). I suspect that, with the same teachers and using the Bible for reference, it would have taken a very sophisticated mind to distinguish between the two types of teaching.

More interestingly, Peltzer posits the differences between an oral and a written literature. She echoes my point made in using the old Berber's recitation to describe joint participation in the retelling of traditional tales and cultural transmission. She writes that it is through "a common pulsation, and with light touches always repeated, that the speaker will communicate the fundamental message of his allocution." It is not the words themselves, she explains, but their passage through the guts of the man telling the story that gives them life (1993:434).

Tahitian poetry reflects the constraints inherited from the oral tradition. Mateata-Allain evokes the importance of words, both intrinsic and contextual, and their relation to the Tahitian mythical representation of the cosmos. "Words and sounds have particular vibrational frequencies that in ancient Tahitian society could offend the gods and the ancestors, resulting in chaos, disease, and devastation." Moreover, and this new dimension has a strong impact of the modern written expression of literature, once those words "were sent up into the cosmos," they could not be retrieved. A new danger now arises when these words are put to paper, because they become "materialized," therefore "increasing prospective social consequences." Finally, when these words appear in print, as modern literature usually demands them to do, they become "tangible and can never be retracted" (2009:6,9).

The European literatures with which we may be more familiar are too far removed from their oral origins to suffer the same moral and cultural ambiguity we find in such societies as Tahiti. We have long ago passed from a former powerful epic medium in which all cultural beliefs were anchored to the personal quests of modern literature. Tahitians have only recently experienced this conversion, usually doing so in a language foreign to their sources.

Today, we are so wedded to the written word that no one recites *La Chanson de Roland* or *Sir Gawain and the Green Knight,* and we must read them in books. Yet, there is in our own societies a renewed interest in the work of storytellers. We find them in schools, libraries, or in festive occasions where people gather. Most Tahitian writers would recognize the value of this storytelling, and would also recognize several precise variations in the oral transmission of texts. For instance, Flora Devantine discriminates between *parau pa'ari* (legend and all that belongs to the oral tradition), *'a'ai, 'a'amu* (legend, fairy tale, story, fable), *pehe* (traditional song or light-hearted ditty). Daniel Margueron specifies that the terminology used for the various oral genres rests on the contents, themes, and tones of the narratives or on the circumstances or their being recited. He discriminates between *parapore* (a different spelling from the one used by Devantine: epic and elegiac poetry), *anau* (complaint), *rauti tami* (harangue), *patautau* (songs for dancing, maing tapa, etc.), *ute* (humorous songs, epigrams), etc. (2002:2).

Devantine created the word *oraliture* to define "a collection of spoken narratives [that involves] a particular way to leave tracks in memories through a voice that imprints and tattoos accounts into the spirit." She sees an evolution between the oral tradition and a modern Tahitian literature in the three words that define these stages: *oralité, oraliture,* and *littérama'ohi,* tracing the path from the oral tradition to the written form of an essentially Tahitian body of work (2009:7).

Tahitian Literary Forms

For European readers, speaking of "Tahitian literature" is not saying anything very specific. While building my bibliography, I had first envisaged a section devoted exclusively to it, with Spitz, Hito, Vaite, Peltzer, among others. But what to do with Loti, Kroepelien, or Melville, who did not seem to belong in the same category as Bougainville or Cook either? Yet, were they not all writing about the beauty of Tahiti, the people's handsome looks, the women's easy seduction, the complexities of the culture? However, the navigators wrote reports and the novelists created fiction—two different categories of works.

Should there be a distinction between Ma'ohi and European writers, which would certainly oppose, say, Spitz and Loti (whom she particularly resents)? So I consulted Mateata-Allain's essays, and found the required conditions for being deemed to be Ma'ohi. A few could be acquired through a long residence in Tahiti, such as, "speaking *reo ma'ohi* or the Franco-Tahitian vernacular; being familiar with insular practices." Others are incontrovertible: having "ancestral and indigenous ties with the *fenua* [homeland]" and sharing a "collective identity rooted in Ma'ohi values." They come from being of the land and definitely exclude aspiring outsiders: "possessing strong links to family, roots, genealogies, oral traditions, and the land," and such links should be maintained in spite of "the rapid changes brought on by civilization"(2009:3).

Obviously, Ma'ohi identity is not something that wishful thinking could achieve. On the other hand, we could not exclude non-Ma'ohi writers from a list of writers *about* Tahiti, particularly when they have so strongly contributed to the formulation of the Tahitian myth, with which Ma'ohi must now contend.

Then, I read a lecture presented in Tahiti in 2002 by Daniel Margueron, which made my life easier by convincing me that all authors should be included alphabetically, without concerning myself about *who* they were and *what* they wrote, since they all wrote about Tahiti, according to their own insights. The broad categories outlined in the lecture are as follows:

1. The *traditional oral literature* in Polynesian language, partially transcribed by missionaries or academics. Composed of odes, elegies, dirges, harangues, epigrams, and others, this literature survived with difficulty the period of evangelization (1800-1820) because it had lost its context and rituals.

2. *Oceanic travel literature*, usually relating adventurous or exotic travel by European and America writers and focusing on the discovery of the "otherness" of Polynesian culture. While we may differentiate between the type of interest shown across the centuries by various writers, such as Bougainville, Banks, or Cook, and Loti, Stevenson, or Melville later, there is probably no such distinction made from a Ma'ohi point of view. All reflect an external view, whatever the century in which it is elaborated. It is also the literature that created and reinforced the Tahitian myth. Bruno Saura has studied some *puta tupuna* (traditional manuscripts) which, he says, include mainly genealogies, lunar calendars, traditions, relations about the contact period, some commentaries about biblical passages. He explains that, while *puta* comes from the English word "book," there are many types, such as *puta tupuma* (book of the ancestors), *puta tumu* (book of the origins), *puta 'a'amu* (book of stories), or *puta parau pa'ari* (book of traditional words). We are once more reminded of the richness of the Tahitian oral tradition.

3. A *neo-Oceanic literature*, written by Occidentals with a good acquaintance of Tahiti or residing there permanently. It consists mostly of works written in French. While the heirs of the exotic literary tradition, these authors are actually inserted into Tahitian life.

4. An *emergent modern Tahitian literature,* started mostly in poetry and later translated into French, with such writers as Henri Hito, Turo Raapoto, Charles Manutahi, and Hubert Brémond.

5. A *modern "tahitianophone" literature*. The last two categories are deemed to be *post-colonial*. It is dominated by writers mostly born between 1940 and 1960, such as Louise Peltzer, Chantal Spitz, Jean-Marc Pambrun, Michou Chaze, Jimmy Ly, and Taaria Walker.

The Literature of Resentment

There is, of course, no such category in the above classification. To call it The Literature of Resentment is perhaps an erroneous representation of its ulterior motive: praising Ma'ohi, celebrating cultural pride, and hoping for future generations to be part of the same culture. "Oh, love of my land," wrote Henri Hiro, "Long live this love, long live!/And may it ever live/And water my native land/So that in their swarming bloom/The children of this soil, children of my land."

Yet, each praise, each expression of faith in the value of the culture, has to be seen as a rejection of the status quo imposed first by the missionaries, then by the French colonial system. The mere fact of saying *Polynésie française* or, in its Tahitian version, *Porinetia farani,* is to impose the burden of colonialism, as neither of those two words evoke Tahitian identity. The first was originally given by a Frenchman to describe the multiple nature of the islands; the second indicates a French claim to possessing some of these islands. It is worth repeating that the words Tahitians use to refer to their land are *Fenua ma'ohi*, which have no connection to the words Polynesia or France.

Tahitian poetry is a long ode to the homeland, an exhortation to remain Ma'ohi. In one of his poems, starting with "I do not understand," Turo Raapoto harangues and berates Ma'ohi for accepting as their own a government imposed by others, when in the past they used to govern themselves. "What does it mean?" he wonders. Feigning incomprehension is a common literary device to stress the magnitude of the notion one obviously rejects.

Similarly, Chantal Spitz compares her people to "well trained monkeys," who have passively accepted their language, their way of thinking, their values, their tastes handed out by others. She asks, forcefully, "Ma'ohi, what have they done to you?/ Ma'ohi, what have you done to yourself?" (2003:8).

In Tahiti, most of the writers are associated with the ever-present nationalist impulse and, during the late 1960s and early 1970s, the men led the way with poets Henri Hito, Turo Raapoto, Charles Manutahi, and Hubert Brémond. But others, as well, berated "those people who came to destroy us now come back en force to educate us," and wonders whether they have lived so long to see this shame, and to see the death of the spirit that gave them their life (Rui a Mapuhi).

Today, many women writers are part of the post-colonialist and "Tahitianophone" group. The most vocal is perhaps Chantal Spitz, often quoted here, in her resentment, both of those who have deceived her people and of her people themselves. She writes, "Ta'aroa's [Ta'aroa Te Tumu: god of the sea who called the world into being] pain is our pain today. People of my Land, People of my belly, Ma'ohi People, Ma'ohi of today. I look upon you now and I do not know you" (2007:20, 162).

[XI]
SHATTERED DREAMS AND MATERENA'S WORLD

IT IS FAIR TO ASSUME that books written by Tahitians offer an insight into what constitutes modern life in Tahiti. Literature has always provided the key to the societies from which they evolve, and it would be difficult to argue that Dickens and Austen, Balzac and Zola do not provide a window into the souls of their contemporaries nor reflect the concerns of their societies. Naturally, I cannot compare the scope of their works with the books I will be examining here. The two authors I chose, Chantal T. Spitz and Célestine Hitiura Vaite, are certainly not prolific writers; on the other hand, they can be seen as exemplars of a nascent Tahitian literature. Interestingly, and in spite of the overt difference they present in style and focus, it soon becomes apparent that they are in fact the two polar extremities of a same axis.

Spitz and Vaite

The cover of Chantal Spitz's book shows the dark, slightly androgynous, and handsome face of a young Tahitian, the eyes almost hidden in gloom. Black spikes of palm tree fronds frame the dark green surface of a lagoon or the sea. The English title claims the island's dreams have been *shattered*; they have been *écrasés* (crushed) in the French one, and either term is perfectly reflected in the timeless sadness of the face.

Next to Spitz's book, the Vaite trilogy jumps out of the cover with childish gaiety. The words, *breadfruit, frangipani, tiare, bloom*, all have strong positive connotations. The drawings could have been made by a talented youngster who only had bright, cheerful crayons handy, and favoured sunny yellow. They represent happy people wearing bright pareus and crowns of flowers, lagoon and mountains, plants in bloom, and a fast sports car to show that the tale is contemporary.

Spitz and Vaite's book covers

Based on the covers and the titles, there is little doubt that we will be entering two different worlds as we read these two authors. This will prove to be true. Yet, it will also prove to be quite false, as both Spitz and Vaite, using a different pen, write about the same Tahiti, albeit a few years apart.

Chantal Spitz, the novelist and post-colonial activist, is acknowledged as one of the authentic voices of Tahiti. She is a controversial writer with strong political opinions always strongly worded. A descendant of the French photographer Charles Spitz (1857-1894), she was born in Papeete in 1954 and now lives in Maeva, on Huahine. She writes in French, the accepted language of modern literature in Tahiti, and when her work is translated into English, it is in order to be disseminated abroad.

She actively contributes to *Littérama'ohi*, a literary review devoted to introducing and celebrating Polynesian writers. In her many interviews and articles, she continues to fight against the stereotypes of the "good native" and the lascivious *vahine*. Her works include *Hombo, transcription d'une biographie* (2002), as well as *L'Ile des rêves écrasés*, (translated into English as *Island of Shattered Dreams* in 2007), and a collection of essays, *Pensées insolentes et inutiles* (2006). In a 2008 presentation in front of the Assemblée de Polynésie, she put forward what she, and others like her, stand for:

> résistance
> résignation
> ni l'un ni l'autre (neither one nor the other)
> et pourtant l'un et l'autre (and yet one and the other)

This ambiguity is expanded in *Island of Shattered Dreams*:

> "What can you say about a people torn apart /
> Caught between two worlds forever divided.
> In one, Yesterday is behind them /
> And Tomorrow ahead of them.
> In the other, Yesterday is ahead of them /
> And Tomorrow, behind them" (2007:89).

Her style is passionate and lyrical, both in verse and prose, and often inspired by the old traditional *ma'ohi* chants. She relates in *Island of Shattered Dreams* the experience of three generations living through several phases of Ma'ohi-European confrontation and co-existence, living as well the most passionate love affairs overseen by Tetuamarama, the moon, goddess of the night and companion of all lovers. The latter's protection helps give an almost mystical dimension to the fateful meeting of souls and bodies for several pairs of lovers as they live through events that are still shaping Tahitian society.

Totally different from Spitz's book, both in style and in approach, is the realistic trilogy we consider in our attempt at discerning through literature the main concerns of Tahitians. Its author, Célestine Vaite, was born in Tahiti in 1966 of a Tahitian mother and French father. She lives in New South Wales, Australia, where she wrote *Breadfruit* (2000), *Frangipani* (2004), and *Tiare in Bloom* (2006). Her style is usually prosaic and often colloquial, and she writes in English. In her case, rather than introducing Tahiti to the world, the translation has the reverse effect of bringing her books home to Tahiti.

With Vaite's unusual position as an expatriated and less overtly political writer, I had to wonder where she belonged in the panoply of Tahitian literature. Was she now merely an Australian outsider writing about her memories of a childhood in Tahiti and, in spite of her roots and regular visits to her homeland, had she perhaps lost touch with the spirit of *Ma'ohi*? I have not seen her name listed among the Tahitian writers of note, but her body of work is perhaps too recent for her to be included by those who make such lists. Or perhaps the list-makers and critics focus on a literature that is more relevant to them and reflects their own political resentment and literary quest.

She writes novels about the daily life, character, comportment, connections of a Tahitian woman, mother, lover, daughter, worker, cousin, listener, and pillar of her small yet far-reaching society. Matarena is woman in all her many guises. That she is of Tahiti and has connections in the other islands is amply illustrated as she goes about her daily chores, reminisces about family history, reflects upon religion, recollects customs, or gives thoughtful guidelines for appropriate behaviour.

The proof that she belongs among Tahitian writers as a singular, yet popular voice is provided by another voice: *vox populi*. Or more exactly, the voice of the students of *l'Université de la Polynésie française,* who gave her twice the *Prix littéraire des étudiants*: in 2004 for *Breadfruit* and in 2006 for *Frangipani*.

A few pages into their books, we realize that each writer has two distinct voices. Spitz writes in French, with Tahitian words interspersed throughout; Vaite writes in English, with the same scattering of Tahitian. However, their two voices are not French/Tahitian for Spitz or English/Tahitian for Vaite, but pertain more to structural needs that call for different styles within their work.

For Spitz, the duality consists in the alternation of poetry and prose. Her poetry is that of the traditional Tahitian epic form, an explosion of the soul that must every so often break away from the realistic confines of modern European novelistic conventions. She appears culturally torn while passing from one to the other, yet at all times, she seems to remain true to her sources.

Vaite, responding to the apparent incongruity for a Tahitian novelist to write in English rather than in French, explains how she proceeds. "I always act out my dialogues… and I talk in French as my character… would. Then I write in English. As for the narrative voice, it comes out directly in English but with a French/Tahitian voice in my head, as if my mother or auntie were telling me the story. Very often I'm translating literally" (2006a:4). This is where the duality of her style comes in.

I am reminded of commercial advertisements that used to appear on Canadian television in the 1960s. Often, the little housewife who did not know better than, say, do her laundry the old-fashioned way, spoke with a strong Québécois accent. Then came the voice of reason/science/experience who would teach her how to use the modern appliances being advertised. That voice was often male but, even more significantly, spoke "Parisian" French: the accent of education and prestige (at the time, for things have definitely changed since then). I sometimes had the same feeling reading Vaite's books. The home-grown voice (which she rehearses in French before translating it) relates to ordinary events and the common occurrences of daily life. The discourse voice (the one she says comes directly to her in English) is the voice of almost bookish knowledge. In it, she deals with history, politics, traditional beliefs, and customs. The two voices superimpose each other at times, and sometimes alternate. Listening to them, we can guess that she is used to *living* as a francophone Tahitian (even in New South Wales) and doling out facts and explaining them as an English-speaker.

A second difference between the two writers is that they position themselves a few decades apart and the political involvement of their characters reflects the concerns of their times. For Spitz, three men (Maevarua, his son Tematua, and his grandson Terii) represent the political stance of their generations because they are involved in World War II and in the opposition to nuclear testing. For Vaite's characters, the memory of these events still exists, but they have become issues of the past. What remains is mostly the malaise of modern Tahiti and, for some secondary characters, an interest in political engagement.

The last significant difference between the two is that Spitz does not make any concessions and presumes that her readers will work at understanding what she is saying. She expects that they will appreciate the historical and political background in which her characters evolve and to which they react forcefully. Vaite, on the other hand, is pleased to escort her readers through the minutiae of daily life and the interpersonal exchanges and conflicts of Materena, her family, and her friends. Since she too must place them in a realistic context, the explanatory voice will then break through to do just that.

History and politics

Vaite's method covers a lot of ground, placing discussions and events in context through a form of "lecture" that helps the reader understand what the discussions and events actually mean beyond the words the participants use. Her lectures often evoke recent Tahitian history, as when, for instance, Ati suddenly becomes interested in politics because his girlfriend has left him to follow a *popa'a légionnaire* back to France. This event is an opportunity for Vaite to explore Oscar Temaru's independent party, which Ati has joined. Ati intends to participate in a rally where all will grab a broom and sweep the roads to symbolize getting rid of the French *popa'a,* "those invaders, those wicked people." By now, Ati has gone beyond his immediate reason for hating the *popa'a,* and is reaching into the past. "When *pai* France needed patriots during the two World Wars, eh, we volunteered, yes, we volunteered to defend *la patrie,* because that's what you do when *la patrie* needs you… *La patrie* called out for help and we responded, and by the thousands, but when it was us who called out *la patrie* did the deaf trick on us." When Ati wants to explain his reasons for joining an independent political movement, Vaite's didactic voice takes over and gives the readers a history lesson about the French government's decision to use the Mururoa Atoll for nuclear testing and the Tahitian people refusal to accept this decision. Her tone is sober and almost academic. At the end of her exposé, Ati once more takes over and concludes, "And those bastards exploded their bomb in our country. Our country!... France gave us money to shut our big mouths… and too many of us accepted, and since then we're all *foutue*. The whole lot of us. *Foutue*" (2006a:173-76). "Screwed" would be an adequate translation for *foutue*.

A few other truths are revealed by Vaite about past land deals done "under private seal," obviously not advantageous to the ancestors, who sometimes traded their land for a few quarts of wine. Vaite does not issue any overt moral condemnation, but Spitz writes on the same topic that "a people loses its soul when it sells its land."

Tetiare, the youngest daughter in Spitz's book, goes further back in the ways of trust and betrayal. "We were duped from the very beginning, we were trusting and naïve." Told by the ministers to close their eyes and pray, the Tahitians did. When they heard the ministers say "Amen, peace be with you," they opened their eyes and saw that their world had been stolen from them. "History is still repeating itself today. We close our eyes and people steal from us," says Tetiare (2007:122).

With Spitz, history and politics are integrally woven into the fabric and structure of the tale, and there is no change in tone, except when she switches from prose to verse, but in both, her voice remains lyrical. The focal points of the text are World War II and nuclear testing in French Polynesia.

About the war, Spitz describes the same political call to arms we saw with Vaite. "The Motherland is in great danger… All the children of our great Nation will rise and defend [the Motherland]," says the French *militaire* come to recruit them. But, being Ma'ohi, the young men do not understand what that "motherland" means. All they really hear is the opportunity to get away and see the world; and they have no idea what the war ("the white man's madness") means (2007:26-27). Of the seventeen who went, only five came back. Tematua is one of them, and never speaks of his experience overseas.

A generation later, the same Motherland has different needs, as she now decides to establish a nuclear missile launch on Ruahine. Upon hearing the news, Tematua tastes blood in his mouth. "A taste of blood of his brothers who died for this glorious Motherland… A taste of the blood of his people's violated soul." The decision to bring nuclear power to the atoll results in a "sudden, uncontrollable explosion," for which no one had been prepared. The decision will bring "profit, envy, poverty, delinquency, prostitution, pollution, exploitation," (2007:73-74). It is feared that it will lead to the destruction of the people, the culture, and the environment. *"Foutue,"* as would say Ati in Vaite's *Breadfruit*.

As the Tahitians are unable to distance themselves from their roots, so are the Europeans. Two generations apart, Charles Williams and Laura Lebrun are caught up in their own positions and backgrounds and cannot escape from their life paths. Charles is a businessman who buys and sells properties, in spite of their being of much symbolic value to his Tahitian daughter Emere and her family; Laura is a technician at the nuclear testing centre her lover Terii has done his utmost to prevent. Both are constrained by their "white" interests.

Connection with the Land

There is a strong attachment in the Spitz's novel to tradition. Much is made, for instance, of the union of the child and the land soon after its birth through the burying of placenta and the *pito* (umbilical cord). Maevarua buries his son's placenta in a place carefully chosen in the family's land. "He opens the belly of the bountiful earth," writes Spitz. Then, he places the placenta in and plants a young *tumu'uru* (breadfruit) on top, before replacing the soil (2007:24). The symbol is evident: the placenta has nourished Tematua inside his mother's belly, and now the breadfruit, feeding on the placenta inside the belly of the earth, will nourish the child throughout his life. In turn, Tematua performs the same ritual when his daughter is born, "the timeless act of uniting his daughter with her Land." He plants for her a *tumu tou* [a tree with medicinal properties], which will heal her throughout her life.

None of Vaite's characters are seen performing the ritual. However, when listing what a true Tahitian must know and do, Materena mentions "planting the child's placenta in the earth along with a tree." She appears to be more concerned with where to be buried, and would prefer to be buried with her mother because "once you're linked with your mother through the umbilical cord, you're linked for eternity" (2006a:57). This is another symbolic way of enunciating the connection between the human mother and Mother Earth through transferring the child's umbilical cord from the former to the latter.

Love and Attachment

Love is paramount in Spitz's novel. Passionate in the extreme, if affects several couples: Toofa and the wealthy Englishman Charles Williams (married to another), parents of Emere (Emily); Teuira and Maevarua, parents of Tematua; Tematua and Emere; their son, Terii, and the Frenchwoman Laura Lebrun. The Tahitian couples experience their love as a natural and forceful event that gives sustenance to their lives; the mixed couples experience it as an equally fateful extension of their respective cultures, but one that is impossible to consume to the fullest. Often occurring at the very first glance, passion is for Spitz's lovers something that swoops into their lives, "intoxicating your bodies with shared pleasures, shattering your souls with your new-found love." They never recover from it.

This is how Tematua expresses his love to Emere. "Within me I store our wild pleasure / Flaming in an endless rainbow. / Not the tiniest hesitation. / Absolute harmony. Complete happiness. / Fusion of bodies. Communion of souls. You were made for me / I was made for you" (2007:52-53).

Vaite's Materena and Pito would probably be embarrassed to speak of love in these terms. The love they feel for each other, which usually remains unexpressed, would appear pedestrian to Spitz's characters. In their world, it is accepted that the way to a woman's heart is through small gestures of affection and that she is moved by signs of tenderness, while a man's heart is reached "through his stomach… or more likely, in Materena's experience, his *moa* [sex]." Plain words indeed, particularly when compared with Spitz', where we find Tetuamarama flooding [Tamatua's] "belly with an unbearable longing for her." Tematua speaks "to the wind and the stars," saying, "My belly cries out its limitless pain." It is only a difference in style and perhaps not such an important one after all, since all it means is that Spitz's Tematua loves Emere and Vaite's Pito loves Materena.

Pito admits to himself that he likes his wife's small ankles and wrists. He gets jealous, and even thinks of leaving briefly when Materena goes dancing one evening with a girlfriend and, deciding she should not be driving home, only comes back the following morning, without explanation. But, even then, and unhappy as he is, he does not go to extremes of passion and grief at the thought of having lost his wife's love.

Love certainly borrows different voices with the two writers. Spitz's characters are so driven by it that it becomes an excess in itself. The sky, the sea, the ancestors, and the gods are all made to witness the way the couples are ruled by their passion.

On another level entirely, Materena, who has become the respected host of a popular women's radio talk show, asks her listeners what is the wildest thing they have ever done for love. "Doing crazy things for love when they couldn't think proper," includes lying to a judge to provide an alibi for a boyfriend, giving up something they enjoyed (sewing, breeding chickens) but that displeased him, driving a boyfriend home without permission or even a licence, etc. No man calls Materena, but the women continue with the list of their past deeds. "Swimming across a shark-infested channel. Waxing. Tattooing his initials on the lower back." Finally, a man calls and it is her daughter Leilani's boyfriend, announcing that he will fly for a few days to Paris where she goes to school to surprise her. Why does he want to do it? "My heart… longing for her so bad." Leilani herself once saw from behind a man of similar built as her boyfriend and she "froze, right there in the middle of the footpath… She was like a coconut tree. And her heart was going bip-bip" (2007:141-43).

So, love exists as powerfully in Vaite's Tahiti as it does in the Tahiti of Spitz. It is not expressed through the same words but, while sounding at times ordinary and prosaic rather than fateful and earth shaking, it equally sustains both sets of characters.

The *Demis*

The authors have much to say about a form of malaise created by the *Popa'a*/*Ma'ohi* combination: the half-caste, the half-breed, or the *demis* as they are usually called in Tahiti. What could be expected from the merging of races and for the children issued of these unions? In Spitz's novel, Tematua is married to Emere, who is half-English. Tematua senses that "his children foreshadow the new world rising on the horizon," a world he sees as divided, but where he hopes each will be able to survive and find happiness. This hope cannot be sustained in the rest of Spitz's novel. For Terii, who must go and live with his grandmother Toofa in Tahiti in order to attend high school, the transition is painful. Through his Tahitian language, he feels connected to his physical surroundings, the "stars, tides, moons, and dreams." Yet, from Toofa (whose daughter Emere was partly brought up as an English girl, her name then shifting back to being Emily), he learns that the white men are "much more intelligent," that they invented "all these extraordinary things, electricity, cars, ships, places." But he knows that Tahitians "had invented a culture too," seemingly dismissed by his Tahitian grandmother (2007:64-5).

Terii and his two sisters, children of Tematua and Emere, are a mixture of two cultures, and as "new children," will never be "whole." Their struggle is "futile" because they no longer know "the language of the belly of their people." Their "minds and spirits" will be attuned to the world of white people, but "their souls will cry out with the pain of their Land and their People. Eternal uprooting of the spirit. Immortal anchoring of the belly" (2007:72). Spitz cannot detach their unhappiness from the treatment received by Tahitians at the hand of the Europeans.

Vaite also writes about the confusion felt by *demis*. One day, on Materena's radio show, the topic accidentally comes up, and she discovers all "these half-caste women confused over their identity and feeling like they were being cut in two." When a woman calls, proudly announcing that her father, a French *tropicalisé*, feels even more Tahitian than she does herself, Materena asks, "Do you think of yourself as French?" After a long hesitation, the caller whispers. "I don't know who I am. I'm so confused about my identity. My father, who is French, acts like he's Tahitian. My mother, who is Tahitian, acts like she is French." She thinks, and sighs again, "Who am I? Half-Tahitian, half-French… but where do I go?" The conversation among Materena's callers then becomes confused, everyone adding their opinion on what in fact constitutes a Tahitian. No one agrees, even if some common elements of the Tahitian identity are included in the list of requirements (speaking the language, fishing and growing taro, living like in the old days, showing respect to old people, remembering and honoring the dead, nurturing the soil and the ocean, being diplomatic with the relatives) (2007: 21-24).

Vaite's callers to Materena's radio show are ordinary women who lack the sophistication of Spitz's characters, yet both address their thoughts to the splitting of their dual identities. There appears to be a social difference between their white father (deemed to be superior by his origins, if not necessarily by his qualities) and their island mother. As well, many of the white fathers, often foreign sailors or Frenchmen doing their military service in Tahiti, were never seen again. And, whereas the local girls they dated were considered to be "sluts" by other Tahitians, no such opprobrium was attached to the *papa'a militaires* involved in the same relationships.

Indeed, the *papa'a father* has often left behind an enviable inheritance. The new generation of half-breed described by Spitz does not share the feeling of uncertainty and the malaise experienced by the half-caste characters in Vaite's books when they reflect upon their identity. Rather, they seem to have fulfilled Tematua's hope that this new breed of mixed children would find their own happiness. Writing about the twenty years that have elapsed since the story told in *Island of Shattered Dreams,* Spitz introduced this new "race of the 'half-bloods,' the demis." It does not matter what sort of "white" the non-Tahitian "half" is composed of, as long as it is white. The *papa'a* fathers or ancestors have endowed them with all sorts of desirable traits: a paler skin, possibly a foreign surname that distinguishes them from other Tahitians, better schools, and, above all, an extraordinary sense of entitlement that distinguishes them even further from the ordinary Ma'ohi around them. Ironically, they reclaim their (half) Tahitian identity when abroad, where the myth prevails, and are found "proudly accepting the name of Ma'ohi, and talking loud and long about the magnificent culture of their ancestors to whom they suddenly lay claim." Once they have returned home, they will once more disown it in favour of their half-white and more profitable identity (2007:149-50).

These are the same *demis* we meet in Michener's *Return to Paradise* ("Polynesia"). Sometime after the second world war, when Papeete only had thirteen thousand residents, Michener describes what must have then been the Tahitian elite, among which stands the far-reaching Bambridge family descended from a "fire-eating English missionary with a gift for carpentry and an eye for native girls," whose descendants had by then a finger firmly implanted in most of the islands' economic pies.

Reading Vaite, on the other hand, that sense of entitlement experienced by the *demis* is not readily apparent, nor do we see the characters drawing any particular benefit from their situation. Materena herself would simply like to meet her father, who she believes had a good relationship with her mother, before the latter resented a remark he made about her cooking and left in a huff. Tom Delors then went back to France before Materena was born.

Spitz makes the distinction between the more ordinary half-breed and the *demis* who dominate Tahitian society, and whom the younger daughter Tetiare describes as "the well-bred, the well-connected, who got their step up on the back of the *Kaina* [ordinary Ma'ohi]." Reminded that she is herself a *demi*, she rejoins Materena's callers on the radio show when she explains, "Being a half-blood is essentially a state of mind that I don't have. A half-blood is nobody, neither white nor Ma'ohi" (2007:150).

The *Vahine* and her Life

Both authors often borrow the voice of their female characters, and an image of their own *vahine* gradually emerges, no longer the seductive *vahine* of the navigators or the tourists' myth, but a flesh and blood, modern Tahitian woman busy with her own everyday concerns, be they the political environment or shopping at the Chinese store.

One of Spitz's well known "rants" ("Rarahu iti e autre moi-même") is against the image of Tahitian women Loti unveiled to Europe, a cliché and an offence to each one of them. When Laura Lebrun first arrived in Tahiti in the early 1960s, she expected to see "lascivious maidens in grass skirts, their hair cascading down their bare breasts." Instead, she met the women in her lover's family, none of whom remotely corresponded to the cliché she was evoking.

Vaite's Materena also plays an active role in reducing the cliché to the normalcy of ordinary life. It would be difficult not to like Materena and not to be reassured by her presence. She has been compared to Mma Precious Ramotswe, the famous owner of the No 1 Ladies' Detective Agency in

Botswana, from a series of books by Alexander McCall Smith. It is almost as though sensible women with a full life, issued from the wisdom of other continents, are deemed to be, if not interchangeable, as least so reminiscent of one another that they must immediately be made to evoke the others. They may merely relate to another myth, that of the wise woman who incorporates the female experience from birth (her own and those she gives) to death (her own and those she accompanies) not only of her race, but of the ages.

None of the women we have met in Spitz or Vaite's books could pretend to incarnate the *vahine* of myth, nor would they acknowledge having much to do with her.

At the centre of Materena's life are Pito, their three children, and her mother Loana, to whom she is very close. She is also proud of her work as a "professional cleaner" at a Frenchwoman's house, and later as the host of a radio talk show. After the doom and gloom that rightly permeate Spitz's *Island of Shattered Dreams,* and the anguish that fills its characters, Vaite's trilogy brings us down to earth and the routine of daily life as ordinary people are likely to lead it.

One of my tests after reading a book pertaining to describe a society, whether a novel or a non-fiction description of its infrastructure, is always to wonder whether I could almost (meaning superficially) function in that society without making basic and horrendous mistakes, or at least understand why things are done the way they are. In the case of Tahiti, it would also mean understanding the values and constraints of family ties, most of which serve to underpin the intricacies of social intercourse, and the deep connection to *fenua*, the land—to the point of fearing dying abroad lest the soul could not find its way home to Tahiti.

Branislaw Malinowski, an important anthropologist of the last century, writing of his ethnographic fieldwork in the Trobriand Islands, explains that while observing and participating in village life, he "acquired the 'feeling' for native good and bad manners." Novels can often serve as a shortcut to obtaining the same feeling through descriptions of scenes, events, and actions, provided that the reader believes in the writer's authenticity. Unable to witness Tahitian life as an ethnographer, I needed to rely on Tahitian "fiction" to review what constitutes appropriate behaviour.

The wisdom and generosity of Materena's culture show particularly when she writes about what to do when a new baby arrives, even before the Welcoming to the World rituals occur. Since we coincidentally had a new arrival in our own family, young Henry, I was particularly sensitive to Materena's advice. First, she writes (in *Breadfruit*), a brand new baby has to be admired for at least fifteen minutes. I am not sure that our little Henry was given fifteen minutes of *undivided* attention by each woman of his family. While the conversation revolved *around* him as a passive participant in the events of his arrival (his mother's long and exhausting labour, the surgical manner of his birth, his father's calmness and exemplary conduct), it was not exclusively *about him*. As well (in *Frangipani*), we are told that when admiring the baby you should not make comparisons with other children, even if these are well meant; presumably, each baby is unique and bears no comparison with anyone else. In other words, babies must be given their dues as admirable individuals and unique new members of the community. As well, when a mother shows you a photograph of her child (whatever the child's age, even in adulthood), you do not just glance at it; you have to look at it for about half a minute, smiling all the while. Finally, kindest of all, you should not try to look your best when visiting a woman who just had a baby, as she is probably not feeling or looking her best herself.

Materena Mahi has no pretension to being a wise woman. She has her rules of conduct, which we naturally accept as being those of Tahiti, simply because they seem to fall so well within what we understand of a society built on connections. To give a small example of this characteristic, a young man had told me that Club Med had to close down on Moorea because the thirty or so families who owned the land on which it was built could not agree on the terms for a new lease when the time came to renew it. As a consequence, two thousand jobs were lost (I do not guarantee the accuracy of the facts, but merely of the story he told me). I asked whether there was any resentment against those who had failed to agree, and his response was "they are all related, they all know each other," his tone implying that my question, which could have been pertinent in my own culture, made little sense to him as the response was self-evident.

What do we see when we read about Matera's daily life? When she is absorbed in her thoughts and walking by the side of the road, if someone toots his horn, she waves without looking up because "it must be a relative." This happens more than once, because she is often absorbed in her thoughts and many people toot their horns at her. In fact, wherever she goes—to the Chinese store or to mass or to visit her mother—she bumps into relatives, particularly cousins (Moeata; Giselle and her new baby; Rita, her favourite cousin; Georgette, the *raerae* (homosexual) DJ; Tepua or *po'o heva-neva,* "head-in-the-clouds; Loma; Lilly; Tapeta; James; or Teva, from Rangiroa; not to mention the cousins' mates (Ramona, Coco)). This is, of course, without mentioning the aunts, the *mamas* and the *memes* (her own mother Loana with her sister Celia, both daughters of meme Kika; Mama Roti, Tito's mother and a rather typical mother-in-law; meme Agathe, or meme Rarahu; mama Teta, the wedding car driver, and so on). "In the space of a half a day, Materena had bumped into six relatives."

The trouble with cousins is that "once they've decided to talk to you, you can't escape. How can you refuse a cousin?" Some overstay their welcome when they drop in, and the polite thing to do in such a case so "nobody's feelings get hurt" is to start sweeping around their feet to indicate that they should "make a disappearance." The mere brush of the broom against the visitor's shoe means "Can you go now?" (2006a:177). An electrician explains that relatives are free to come and go as they please, using the telephone, watching the television, drinking the Coca Cola, using the washing machine, or even borrowing clothes. Locking your door is an insult. He concludes, "The problem with Tahitian people is that we have too many relatives who don't have a job." Then, thinking the problem again, "and just too many relatives."

On the other hand, relatives can do small kindnesses, give you a ride in their truck, or let you have bits of carpeting remnants or some extra tiles if you want to remodel your bathroom. There are other advantages as well in having these close relationships created by family connections: there is always a willing ear and a sensible piece of advice available. "Tahitians who pour out their troubles to a psychiatrist don't exist—we talk to a favourite cousin," says Materena, who adds, "or to the priest." But, judging from her daily life, she is more likely to encounter a sympathetic cousin along the way than to march off to the church to consult the priest.

Exchanges of services within the family are expected. However, when the service provided is of some consequence, there are implications attached to accepting it. The French sociologist Marcel Mauss showed in his seminal work, *The Gift,* that these exchanges between people, sometimes seen as gifts, are often the cement that keeps societies cohesive. They are never "free" in that they always assume some form of reciprocity. The association between the objects (or services) given and the people who exchange them creates a bond imbued with spiritual dimensions. Materena, while issuing her rules of conduct, provides an example that must occur frequently in a society that relies on its connections to forge ahead. She explains that if you use your connections to find a job for a relative, that relative must stay in the job until he retires. That is the price he must pay for making you use your influence. Behaving otherwise would show disrespect and make you lose face (2006a:303). This probably common situation is nevertheless a complex one, involving two different types of "gifts": the one you give your relative, and the one your connection gives you. The implication is either that a new exchange will take place to balance out the value of the gift, or that the gift comes as a payment for a previous favour, or that one person will remain indebted to another. When a society relies on an exchange of gifts, advices, or services to strengthen its ties, appreciating services at their true value is what keeps the system alive and well.

It is not always possible to give something of much financial value. Wisely, Materena explains that it does not matter because "when you give a small thing the right way, it becomes a big thing," which confirms the spiritual element that is part of gift giving and receiving. The need for reciprocity is known to be important, and only ignorant people offer nothing for something. "You just don't eat at someone's table for twenty-three days [as an Australian guest did] and give nothing in return."

The inter-connectedness of Maretena's relatives, living in a small place, united by ties of blood and custom, makes the Western reader aware of his own disconnection. At the same time, most cultures' rules of courtesy are the same: they strive to foster group cohesiveness while ensuring appropriate individual privacy and comfort. Their outstanding reason for being is based on the respect due to others, and Tahitian etiquette is true to this spirit.

Materena, for instance, explains that, given the openness of Tahitian houses, it is important when visiting someone to announce yourself while standing away from the house ("they might be doing something of a private nature")—a rule which Mama Roti Tehana (Pito's mother) often fails to respect, forcing Materena to entertain her. Discretion is prized, and "women's talk is secret" is another rule of conduct that protects the individual. This is not an exclusive female duty, as we also learn that what transpires between a man, his cousins, and his *copains,* is equally for their ears only. Certainly, Materena does not always know what Pito is up to. Whatever your sex, if someone tells you not to "put the news on the coconut radio [the gossip line], you're supposed to forget the news as soon as someone tells you about it."

Respect for the other person is shown in many subtle ways. For instance when someone is telling a story, "it is best to act like you don't know the rest of it." With similar kindness, you "let them get on with their story" rather than interrupt with your own examples, and you could also prompt them by "feeding them questions." Our etiquette books make similar recommendations, but in a society that prides itself on excelling in the oral arts, it is particularly important to respect storytelling as an important social skill.

Respect for the elderly is advised. "You don't yell at the mamas," is Materena's advice. However, this rule of conduct is made somewhat more realistic: "Only the papas can yell at the mamas, not the young people." Another feature of everyday Tahitian life is "presentation" and it is important to be seen looking your best. For instance, you should always wear your best clothes when going to the school office, as your reputation and the way your child will be rated depend much of your presentation.

I would not presume to reduce Materena's wisdom to a few sayings, but they indicate the delicacy of balancing the pros and cons of propinquity and being part of a large family. It is not always easy because some members of that extended family may have many failings, yet the done thing is to accept these failings, attempt to behave appropriately, and make the most of what you cannot avoid.

Could we say, after reading *Island of Shattered Dreams,* that we might be able to function in the culture described by Spitz? It is true that Terii's family learns to accept and even love Laura Lebrun, the Frenchwoman full of preconceived ideas and biases when she first arrives, but they only do so because of his love for her and her love for him. Others refuse to have anything to do with her, as she attempts to discover for herself how the islanders live in their village near the nuclear testing centre where she works.

A westerner with no first-hand experience of the ways of life in Ruhaine, yet desirous to avoid social pitfalls, could not rely solely on Spitz's book to form a concrete idea of appropriate everyday behaviour. This is particularly true since the book takes place in the charged atmosphere of the 1960s, in the middle of the political turmoil created by France's decision to switch nuclear testing from Algeria to Polynesia. On the other hand, we are introduced throughout the book to the historical bedrock on which Tahiti's contemporary society is built.

Vaite is far more approachable than Spitz for the reader only interested in seeing how people behave and conduct themselves in the normal way of life. Yet, the deep-seated resentment expressed by the latter writer is a necessary background to understand the deep fissure that exists between the *papa'a* and the *ma'ohi* cultures. The following example does not actually come from Spitz but from the usually more temperate Vaite.

In *Breadfruit* (2006a:277-79), we see a terrible misunderstanding happening. Materena's cousin, Tepua, agreed to have her baby adopted by Madame Pietre, a Frenchwoman for whom she worked. Tepua had understood (encouraged in doing so by Jacqueline Pietre herself) that the adoption would be according to the usual Tahitian tradition, where the mother continues to be an integral part of the child's upbringing, if she so chooses; even if she does not wish such constant intimacy with the child, she will nevertheless continue being connected with it. Madame Pietre, on the other hand, had quickly switched her own understanding of the situation by complying with French custom; the child was now hers, and Tepua was no longer welcome in her house. The uncomprehending and distraught Tepua climbed Madame Pietre's fence to see her baby. She was then arrested and accused of robbery. The French gendarme agreed that stealing was a serious problem in Tahiti. "The problem with those people is that they mistake the verb *stealing* for the verb *borrowing*." The choice of words is ironically appropriate here because, according to Tahitian tradition, Madame Pietre was actually borrowing Tepua's baby, while according to French custom, Tepua, now excluded from her maternal role, is deemed to be stealing Madame Pietre's baby.

In *Tiare in Bloom* Vaite's lecturing voice explains the nature of *fa'amu* ("to feed"), the traditional Tahitian adoption intended to help the mother while "she gets herself together." The *fa'amu*-adoptive parents will love the baby and tends to its needs, but it is understood that it is not their baby and still belongs to its mother. She may then come back, thank them profoundly, and take her baby with her. If she does not come back, the baby's father is then next in line for looking after it. Should he fail to do so, the *fa'amu* parents might then decide to become the child's permanent parents. We see throughout the literature many references to the adoption system practiced in Tahiti, throughout the centuries, and it is important to understand its nature. Tepua and Madame Pietre had both suffered terribly for interpreting a different definition of adoption.[25]

After reading these novels, I wondered what I had learned from them of Tahiti. The authors' interests converge throughout, even if their styles could not be more different. What lessons could be drawn from them? Spitz's story of the lives of her three generations is enclosed between a Prologue and an Epilogue. Those epitomize her two voices. The Prologue starts with "It is a magical moment." The deep voice of Tematua is then heard, the "voice of the eternal Land. Voice of the immortal Fathers," and comes from "the ages of man." What the voice will do is narrate the history of the people of Tahiti, in both verse and prose; the latter's style almost unrecognizable from the former. The Prologue then ends with Tematua's death and burial, and the last sentence, "The sacred birds of land and sea folded their wings to mourn."

The Epilogue, on the other hand, adopts the factual voice that Spitz herself uses in her political writings. It describes what happened in the twenty years following her tale of the various lovers' tribulations. Laura's work at the nuclear testing station is a thing of the past. Things have moved at a mind-boggling pace, Huahine going in two decades through changes that took two millennia to achieve in France. The Epilogue starts with, "Twenty years later, the upheaval is irreversible, in the landscape and the economy as well as people's attitude." These have been "uncertain and eventful times on Ruahine, a season of latent violence amongst a despairing people who have lost their way." Yet, we are left at the end with an alarming prophetic vision: "There's nothing more dangerous than a colonized people standing tall" (2007:156).

We then assume that Vaite's Materena, Pito, and their family now live in this new world described by Spitz. Yet, on the whole, they do not appear to despair, even if they sometimes lose their way, which they do in the manner of those who have inherited a different world from that of their recent ancestors. It is not that they lack the depth of Spitz's characters; it is more that they belong to two different ways of life and two entirely different modes of expression.

Those two authors, who at first glance appeared so different, simply offer two complementary sides of Tahitian culture today. Paramount to the rules of behaviour established in Vaite's trilogy is the sense of connection that must be maintained within the family or, given the extended nature of the family, the same community. Materena, in her many functions as a responsible adult, an extended family member, and a worker with a serious job, *knows* how to behave. Through her, we also witness the concerns of this community built on acknowledged connections and intent on maintaining its web-like exchanges of duties, favours, kindness, hospitality, and support. We also learn how the old ways are known and respected, and how history has its own living memory.

Spitz's characters also know how to behave but, more importantly, they know how to think, how to remember, how to act according to their conscience, their principles, and the requirements of their Land.

The two visions are complementary, the two sides of the same coin. Lyrical or prosaic, they reflect in their totality the past and the present; their conflicts and the compromises that survival requires. They only diverge profoundly when they project that survival into the future. Or rather, when we, the readers, project it for them, as neither specifically do so in these books. But, knowing the characters and the environment, it is not difficult to assume that Spitz's younger generation faces a grim future unless the past can somehow be retrieved and given life again, while Vaite's younger generation, educated and seemingly at peace with itself, looks forward to a profitable modern life.

[XII]
TOURISM AND NOA NOA

ARE THERE TOURISTS IN GAUGUIN'S *Noa Noa,* the pristine, "fragrant" island? Of course, and rightly so, since he was among those who fed the innocent dream to his countrymen first, then to the rest of the world. After the navigators, the missionaries, the whalers, the traders, the planters, one would have expected their arrival as the next logical step of European-driven development. The earlier arrivals had reaped many benefits from the islands, including international reputation from scientific discoveries, conversion of souls, wealth built on whales or commerce. What of the tourists?

I bought in Raiatea the usual postcards to send home. Reproductions of Gauguin's Tahitian paintings for my artistic friends. Brilliantly multicoloured fish and blue water under the sun for our son, who hopes one day to sail his catamaran to Tahiti, and for a friend, who dives in the Maldives. For my anthropologically minded friends some *tapa* reproductions, guaranteed to be "Authentic Tapa – Traditional designs of French Polynesia" in English, French, and Chinese. Fiercely tattooed warriors from the Marquesas, spears at the ready, for our two youngest grandsons, hoping to impress them with their grandparents' adventurous spirit. Finally, for some young men among my acquaintances, I also bought the usual photographs of smiling and beautiful young *vahines,* joyously doing a variety of native-like things that required them to be half naked. I saved the most handsome one for the physiotherapist who had worked for nine months on my fractured joints and who deserved to see youthful physical beauty at its best. However, I did not buy the other postcards of garish sunsets pouring orange light onto unlikely palm trees. I may have thought these were even more "touristy" than the ones I bought.

Picture postcards used to be far more popular than they are now when we carry our own individual means of taking pictures and videos and circulating them back home. Almost as soon as photography became an accepted medium of advertising, postcards depicting exotic places were circulated in the millions. It was the easiest way of seeing the world, and they were intended to meet the expectations of potential visitors to a strange land, whether the city next door or the antipodes. To do so, postcards represented symbolic and evocative markers (the Eiffel Tower in Paris, the pyramids in Egypt, the Taj Mahal in Agra, the *vahine* in Tahiti). There is notably one of a beautiful Hawaiian girl, taken circa 1890, now circulated by the Bishop Museum. She looks no more than fifteen years old, with luxuriant wavy black hair, a *lei* around her neck, and revealing one tiny, perfect breast. She looks cold and distant, beautifully inaccessible. At the time, she was expected to evoke the desirability of Polynesian women, but today, looking at her, it is difficult not to think of the drunken whalers invading her island.

Older postcards from Tahiti often depicted scenes and places showing the progress of civilization in the form of various streets, markets, and so on, where French influence could be featured at its best (I remember seeing similar ones from Senegal in my childhood). But their stock in trade was the beautiful *vahine*, the one everyone had been dreaming of since Bougainville, the main initiator of the myth. They were understood to represent the real Tahitian Beauties (the name of a series, actually). "The postcard… becomes the poor man's phantasm: for a few pennies [it displays] racks full of dreams… [it offers] a pseudo-knowledge of the colony," writes Malek Alloula, who compares it to a seabird producing guano, and concludes that the postcard is, for the times, "the fertilizer of the colonial vision" (1986:4).

Miriam Kahn, who describes several of these postcards, notes that a French stamp was issued in 1913 from the *Etablissements Français d'Océanie (EFO)* representing a Tahitian woman with a crown of flowers and wearing a hibiscus behind her ear (2011:55). What better means of circulation for a cliché than being affixed on letters reaching the whole planet, or even being collected worldwide in philatelists' albums?

Similarly, the paper wrapping the small gifts we buy in Moorea or Raiatea is printed on ocean blue background, showing a stylized map of the best known islands (Tahiti, Moorea, Bora Bora, Raiatea, Huahine), a generic tiki, a black pearl in its shell, a prominent *vahine* (the largest picture on the paper) wearing a crown of flowers with her arm hiding her naked torso, another *vahine* in an outrigger in the background, and *"Souvenir de Tahiti – de Bora Bora – de Moorea"* printed across. In other words, all the expected "hooks." It travels back home with the tourists who purchased their gifts and conveys the same message: welcome, ocean, outrigger canoes, islands, palm trees, sunshine, and beautiful half-naked girls.

Tourists are sought after and come in ever increasing numbers. They are the people who often unwillingly debase the sanctity of revered sites or offend local customs; the curious and often well meaning strangers who gape obediently but with little understanding at what guide books tell them are landmarks of historical or cultural importance. They are also the well heeled travellers who can spend lavishly. All over the world, tourism offices try to lure them, in spite of the deleterious effects their presence often has in some countries. The lifestyles they evoke are barely imaginable by local standards and can be a source of envy and resentment; their unthinking largess encourages begging, particularly by children; their lack of subtlety irritates; their loud behaviour, where discretion may be prized, displeases. They are the innocents abroad whose so-called innocence is synonym with ignorance, the babes in the wood whose naivety calls for conmen to draw them into fictitious deals and tours, or perhaps even worse. They can also be overbearing, loud, and impatient, spending and drinking too much, and often being an unbearably discordant sight for resentful residents. Yet, in many places, and particularly Tahiti, their presence is economically essential, and governmental and touristic organizations are set up to seduce them into coming in even greater numbers.

Their motives for travelling are varied. Some may be genuinely interested in the countries they visit, having widely read about them and being well prepared for their museum and monument visits. They may also have a competitive streak and believe that the more countries they have seen, the more interesting they are themselves. I think of the Frenchwoman in Spain, loudly exclaiming to her husband that the *Transparente* above the altar in the Toledo cathedral reminded her of Râ's altar in Egypt, because in both cases the clever architectural trickery of a hole on the wall allowed the sun to shine through at the peak of the religious ceremony. Spain and Egypt compounded each other in her vision of the world outside France. We all do it, constantly reminded by new visions of old memories. For many travellers, the world may be a succession of vignettes related by a single outstanding feature, or even by the travellers' own perception of it. More to the point, it is seldom a place where they can have much enlightened contact with the local people.

Whoever those tourists are, and however individually resented some may be, they are usually seen as a blessing to the local economy. They sleep in hotels, eat in restaurants, buy souvenirs, attend shows, and take tours. The whole community benefits from their presence. Tourism is at the core of an increasing competitive market, where climate, beaches, diving, archaeology, museums, sporting or cultural activities, advantageous exchange rate, and local attractions compete for the tourists' choice of a destination. All are presented to them in the hyperbolic language of tourism, and, in a country such as Tahiti, where tourists particularly wish to confirm their illusions, it is important that the promotion be adequate to their expectations.

The competition is fierce among countries whose economy relies heavily on tourism, and all shamelessly flaunt their assets. Among her competitors, Tahiti is particularly favoured as the land of mythical beauty. The Boomerang Tours brochure from my corner travel agency advertises the island as an ideal retreat from the tedium of everyday life. It is difficult to resist the photographs of pristine aquamarine pools, cerulean skies, and smiling tattooed attendants wearing floral garlands, all intended to charm the tired and sun-deprived westerner. All one wants to do is lie down on one of those inviting chaises longues, sip cool drinks, and forget the rest of the world. But Tahiti is not the only island to offer the promise of such delights.

Danielsson believed that tourists going to Tahiti were not the ordinary kind. They were not particularly interested in visiting monuments (and only the *tiki* and the *marae*, many of them unrestored, could be described as such), but rather sought "the lost paradise," based on reports of the "natives' free and merry love life" in the South Seas. There is indeed little subtlety wasted in advertising the connection between dream and reality, and any reference to the *Bounty* will serve to evoke the earthly delights experienced by Bligh's men—such as naming a new resort "The Brando" built on Tetiaroa, the island on which the actor settled with his Tahitian co-star. Even if overwater bungalows are banned and the Tahitian style is respected, the ultra-luxurious resort and its grotesquely inflated prices have little to do with the island they attempt to evoke.

Chasing after the myth is the avowed purpose of Tahitian tourism, as shown in the description of the new resort Miriam Kahn read in Papeete newspaper article in 2001, which starts with Beaudelaire's words. "Here everything is tranquility and beauty, luxury and voluptuousness. This is the Polynesia of New Cythera, the myth transformed into reality. This is nothing like Disneyland or some other American-style dream machine." It continues, "There, Polynesian culture is the producer of this legendary Rousseau-like harmony. It is far from an industrialized tourism trap" (2011:138). *Not* Disneyland… *Not* an industrialized tourist trap… Perhaps the lady doth protest a little too much about her exclusive connection with both Bougainville and Rousseau, when tourist revenue is clearly the *sine qua non* of the whole affair.

My own travel brochure makes the same connection and is equally hard to resist. "French Polynesia. Your dream can come true! Just the word itself conjures up images of palm trees, the warm scent of flowers, black-eyed beauties in swaying grass skirts and breathtaking turquoise lagoons lapping on white sand beaches. For hundreds of years, explorers in their sailing ships brought back tales of a 'paradise on earth' where the visual blend with the sensual to create a Garden of Eden in the South Sea."

It is indeed this lost paradise that most tourists are still seeking in Tahiti. However, since the 1950s, the time of Danielsson's writing, tourists have changed. Their numbers have increased enormously, as have the costs of indulging in the carefree island experience they expect to find. Often, what they also seek is privacy, which their comparative (and sometimes absolute) wealth can easily provide. For many, if not most, privacy will mean, ironically, an almost complete isolation from the natives who once inspired their vision of South Seas' delights. The ones they will see are only those intended to illustrate aspects of the culture that created the myth. Sadly, as Moran writes in *Beyond the Coral Sea*, "The 'noble savage' of the Enlightenment had evolved through the 'ignoble savage' of the nineteenth century to the present-day 'vanishing savage' on cultural display for tourist camera" (2003:328).

Competin' for the Yankee Dollar

The inviting promises made in the 1945 Andrews Sisters hit, "Rum and Coca Cola," ("They make you feel so very glad… Guarantee you one real good time… Native girls all dance and smile… Make every day like New Year's Eve"), would be frowned upon today. They refer to another place (Trinidad) and another time (World War II), where women were "workin' for the Yankee dollar." We saw an echo of these words in French Polynesia, with James Michener, where the young women used to go to in search of the good life with the American GIs: "Canned food, Jeep rides. Trips in the airplane. Whole cartons of cigarettes." All this speaks to a mood today's advertisements would not dare to evoke in the terms proposed by the song. Even in 1945, this celebration of the good times to be had in Trinidad was already betraying the intention of the original calypso to denounce the perverting influence of the American troops on the island women. The Andrews Sisters somehow lost the message, but the lure of the island persists.

Tahiti is certainly no exception in realizing that her economic prosperity must rely on tourism and that she has many rivals also boasting of cerulean skies, inviting waves, and sandy beaches—also hungry for the American dollar. I think particularly of two island states in the Indian Ocean, the Maldives and the Seychelles, both blessed with the same attractions offered by French Polynesia, including the beautiful scenery, excellent diving, generally attractive climate, and exotic vegetation. Significantly, all they lack in the comparison with Tahiti is the flaunted presence of beautiful and approachable women.

Comparing the Maldives and the Seychelles to Tahiti was an easy choice. First, they boast about their uniqueness, while actually offering the same beautiful environment, enjoyable activities, and comfortable accommodations as the others. Second, my friend Patricia Fung has been diving in the Maldives for several years, and I know a little of the Seychelles, so we can both go beyond the travel posters in our appreciation of their facilities.

All these islands claim to be private and romantic spots, ideal for honeymoons. I was surprised twenty years ago at the Victoria Airport on Mahé to see all signs not only written in the three official languages—English, *Kreol Seselwa (Créole Sheychellois)*, and French—but also in Italian, until I found out there was a direct flight from Rome intended for honeymooning couples, the objects of intense touristic promotion. Tahiti's particularly claim in this context is that she is already known as "the island of love."

It is indeed a tough competition, and every image counts. Here is a *Lonely Planet* description of the Maldives, where tourism started in earnest in the 1970s: "So superior are its beaches, so cobalt blue its waters and so warm its welcome that the country has become a byword for paradise whether it be for honeymooners, sun worshippers, or divers." One imagines natives welcoming visitors and sharing with them the incomparable beauty of their islands. However, *Lonely Planet* also points out that "Outside the service industry, Malé [the capital city, also the site of the international airport] is the only location where the foreign and domestic populations are likely to interact. The tourist resorts are not on islands where the natives live, and casual contacts between the two groups are discouraged." This enforced isolation is all the more unfortunate that the Maldives have an ancient history and culture worth investigating.

The Maldives, a country where Islam is strong, has decided that half naked tourists basking on beaches must remain in de facto exile. The tourists benefit from the sun, beaches, diving, catering facilities; the government gets the revenue. There is no apparent desire to get to know the other party and it seems to work to everyone's satisfaction. The economy-boosting tourism has become in the last thirty years the most important source of income, displacing fisheries as the former economic mainstay. Even dawning extremist currents do not seem to affect the visitors' return, further demonstrating their satisfaction with the status quo.

Pat Fung comments on her experience of visiting the Maldives. "The setting is wonderful and would be even more wonderful if the world never changed and islanders had continued to live what may have seemed an idyllic life. However, this would have been possible given contact with the rest of the world and, of course, life was probably not so idyllic, with lack of rain, too much rain, severe tropical storms…"

She carefully states that her comments exclusively apply to what she and her husband have experienced on their three-week visits to Maayafushi (four times), Biyadoo (three times), Velidhu (twice), and Filyitheo (once), all islands they chose because they were small and lacked "night life." There, she explains, the staff were mainly from Bangladesh, the pay being generally too low for Maldivians. They only go back home once a year but are able to save up enough money to build a house when they return. Other workers are local or come from Sri Lanka and live in their own compound, grouped by nationality. They prefer the arrangement because of the language and food differences, and this area is off-limit to the guests. There are friendly exchanges with guests in the dining room and in the bar, and the staff occasionally make up numbers with the guests for volley-ball matches.

When visiting such places, I have also noticed that friendly staff become somewhat aloof when met off duty and that it is impossible to go beyond a light-hearted superficiality. Why should it be otherwise, unless we flatter ourselves that we inspire genuine friendship among those hired to serve us for a few weeks, when everything separates us—status, culture, race, religion, wealth, prospects, education, etc.

Most touristic resorts or hotels have a weekly evening on which local folklore is presented. It is a standard practice: costumed dancers and musicians perform, the buffet displays local dishes, and drinks are abundant. They are pleasant events for the tourists, who are usually friendly and show appreciation to the performers. How much cultural accuracy is to be found there is anyone's guess. Asked about these folkloric shows for tourists in the Maldives, Pat replied, "Local Maldivian culture seems to centre, as one would expect, on fishermen's ritual songs and dances. It is usual for each island to put on a special evening each week, when the cooking is 'local' and staff perform 'local' songs, dances, and explain a little bit about the 'old' way of life… One suspects that the customs may be kept alive for the sake of the guests. On the other hand, the local (i.e. Maldivian) staff seem genuinely proud and pleased to see us appreciate both the food and performances. I'm not sure what the Bangladeshi staff think, as they are usually the ones doing the cooking and are sometimes even brought into the performances to make up numbers."

What of the Seychelles? As in the Maldives, tourism became an important economic resource in the 1970s, particularly with the construction of an international airport. There is a far-reaching connection with Tahiti in that Bougainville also came to the Seychelles during his circumnavigation. He even named both the Seychelles (after Monsieur de Séchelles, a friend of his family) and the island of Praslin (after the Duke of Praslin, his patron). However, beautiful as the islands are, Bougainville did not gush, and the sparse and disparate population would never have prompted him to think of it as another new Cythera. The islands were mostly uninhabited until two and half centuries ago, when the first escaped slaves, Europeans adventurers, and Indian traders started settling there. The islands never possessed the rich culture Bougainville found in Tahiti.

I need no travel brochure to know how beautiful the Seychelles are, at least the islands I have seen, Mahé, Praslin, and La Digue. Yet, I saw no tourists frequenting Victoria, the capital city, poking their noses into shops, stopping to take pictures. There were a small number of Europeans in town and on some of the islands, but they were mostly consultants, like me, and expats (usually consultants with a longer contract). The tourists were mostly to be found on La Digue, particularly known for its beaches, and where most of the facilities were designed to serve them; others may have visited the *Vallée de Mai* National Park on Praslin.

The Italian honeymooners were not usually seen walking the streets of Victoria, the capital city on Mahé, because they spent most of their time sunbathing and sipping drinks on luxury hotel beaches, such as the Barbarons Club or the Coral Sands Hotel in Beau Vallon; being taken by hotel speedboats to their special diving spots; or enjoying dinner at their hotels, far more sophisticated than Victoria's few eateries. They would have had little reason to visit the town, where everything was sensibly ordained but not particularly attractive, apart from a few samples of old colonial architecture. Certainly, there was nothing there worth the trip from Europe, except for the natural beauty of the islands, best enjoyed away from towns.

Where tourism really pays off in the Seychelles is with the very rich, and the small, private islands they are known to purchase, and it has little to do with the Italian honeymooners and the few ordinary tourists staying in the beach hotels. International interests have been building exclusive and luxurious resorts and retreats for millionaires all over the smaller islands. This type of tourism does not promote much interaction with the native Seychellois for the very select few who can afford the superlative costs of total privacy. Contacts are mostly with nearly invisible servants, intent on not spoiling paradise on earth for these exclusive tourists.

While on La Digue, I was invited to attend an evening of songs and dances meant for the local people. Young girls, dressed in "formals," sang loudly tunes in which I recognized the odd French folk song reworked almost beyond recognition in their Creole versions—a nod towards the French possession of the islands from 1756 and 1814 and the affinities between *Kreol Seselwa* and the French language. There was also a short demonstration on stage of the *sega*, the local dance one is supposed to "roll." The show was much appreciated by the audience, mostly workers involved in the various touristic services on that small island. The interminable length of the show, the endless repetitions, and the extremely amateurish performances would not have appealed to tourists, and neither was it intended for them. But times have changed, and a recent documentary, twenty years after my visit, showed "folkloric" dancing on the streets of Victoria by performers wearing presumably traditional costumes. They belonged to dancing groups and schools, some dancers, in spite of their very young age, even "rolling" the *sega*. I could not see whether there were tourists among the few onlookers.

In Tahiti, the tourists are not excluded from everyday island life, and neither does the government restrict tourists from visiting any island, as in the Maldives. Yet, tourism in Tahiti faces the same separation from the realities of ordinary island life as it does in the Maldives, Seychelles, and any other supposedly idyllic retreat in the tropics. Driving around Tahiti, Bora Bora, or Moorea, a familiar sight is that of series of "native" looking huts built over the water and at some distance from any village or urban centre—too far to walk to the next town to face a little reality check. Life is made to appear luxuriously self-sufficient in this gathering of huts. All the tourist wants is supposed to be available there: the wonderful blue lagoon, comfort, service, food and drinks, as well as entertainment brought to the resorts in the form of dancers and drummers well aware of the visitors' tastes. In fact, they are so well aware of their expectations that some tourists in the audience may even feel patronized by being so easily outguessed.

Typical Resort (Moorea)

Were tourists inclined to poke around freely, they would see many contradictions to the official picture that is flaunted. The shack towns of bidonvilles exist; people are poor and there are slums; there is also a backcountry with few facilities. The sharp jolt of seeing another and more realistic aspect of local life is probably deemed to be too upsetting for tourists, who are spared the sight. Tourists know that their experience on these islands is not the reality of life for the natives. But they are on holiday, they want to enjoy themselves, so they go along with the fiction sold to them at substantial cost.

Resorts advertised as being in Papeete are in fact several miles out of town, with little or no easy access to transportation outside the resort. The point is well made and illustrated (in her chapter "In the Cocoon") by Miriam Kahn who had many opportunities to examine the dissociation between the place where the tourists are, how they perceive it, and how they live their experience. There is usually little infringing of real life into those heavenly touristic spots. Most tours are organized to show only what will enhance and reinforce the myth, and the only natives they meet are the bus driver and the guide tour, both well versed in maintaining the fiction. This type of tourism is usually described as travelling in a "bubble" or in a "cocoon" and mostly focus on the planned entertainment and companionship resting within the bubble itself. Thus, tame tourists mostly see what is shown to them and mostly consort with one another.

It is not easy to step away from the buffering bubble and being totally taken in charge if you have accepted to be isolated from the local foreign environment. I once attended a large international conference in Bangkok during which I only had one free afternoon to step away from the presentations and the organized activities (seminars and lectures, but also banquets, folkloric representations, tours, and so on), and I felt totally lost while abandoned to my own devices. I do not think upon that week in Thailand as "travel" or "tourism" but as an out-of-reality experience that only brought me a little understanding of the culture of large international conferences. Similarly, visiting holiday spots renowned for their beauty and their escapist lifestyle mostly gives an insight into the systematic exploitation of touristic come-ons, rather than into the land visited and the people who inhabit it.

This type of tourism emerged in the middle of the nineteenth century, when Thomas Cook, a Baptist lay preacher, organized the first excursion to the Midlands for members of the Temperance Society. It has since become much more sophisticated, but the same principle remains: protecting the tourist from the pernicious influence of the locals, and even more likely nowadays, from the sad reality of the natives' daily life that may upset the delicacies of the tourists used to better standards of living.

One is not likely to find among these tame tourists the modern equivalent of Lord Curzon, reading between two hundred and three hundred books written over the past five centuries in European languages and Persian before setting forth to Persia in 1889. Some may have read Loti; others may be familiar with writers of South Seas tales; most have watched *Mutiny on the Bounty*; all must have seen Gauguin's paintings. A few may even have read Cook or Bougainville's reports. It matters little how specific their knowledge of the islands is, or how they came by it, because they probably all carry the same baggage of the Tahitian myth. Once in Tahiti, it is easy to allow the myth to take over and overshadow the reality of Tahiti in the twenty-first century.

While in the bubble, there is actually no particular need to meet any "ordinary" Tahitian. Contact with the local people can be reduced to the hotel and resort staff, and to the entertainers and dancers who offer a version of what most tourists assume to be Tahitian culture. A popular draw, the Tiki Village in Moorea, presents a "big sunset show with dancing and drumming." For those with a taste for something more exotic and willing to pay CFP 150,000 [very roughly $1,600], "a 'royal' Tahitian wedding can be arranged at the village (bring your own partner; same sex couples welcome). The ceremony lasts two hours, from 4 p.m. to sunset. The bridegroom arrives by canoe and the newlyweds are carried around in a procession by four 'warriors.' Otherwise, there is a less extravagant 'princely' wedding for CFP 120,000 [equally roughly $1,300], photos included." The guidebook reminds us that such weddings are not legally binding (Stanley, 2011:98).

We walked through the Tiki Village on an ordinary day, without any show or craft demonstration scheduled. There were few people, and the replica of Gauguin's house was closed. Half a dozen shops were open, selling the usual jewelry, carvings, and pareus. Created by a Frenchman, Olivier Briac, as a tourist attraction and commercial outlet, the Tiki Village seems appreciated by visitors. Online advertisements read: "Meet the Polynesia of your dreams, far from sterile environments, in an idyllic setting" or, "Visitors experience the traditions and lifestyle of an old Tahitian village through demonstrations of weaving, stone carving, tattooing and painting." Visitors' comments show them to be, on the whole, well pleased with the experience. For some, it seems an accurate recreation of the past. "You really get to see the true Polynesian culture. They do all live here and you get to see their everyday living." For a few others, the food was average and the dancing went on for too long.

We were greeted along the way by a man to whom I said, *"Bonjour."* He corrected me, faking sternness. "Not *bonjour*. Here, we say *ia orana."* Then I met another man, wearing a necklace of shells and what looked like large and slightly wrinkled amber beads. I asked what they were. He said "Cows' balls!" "Mummified?" I asked. He and the woman beside him laughed and he explained they were the larger seeds of a small tree, taking me by the hand to show me one of those trees nearby. There is another smaller seed inside from which they extract an oil used for many health purposes.[26]

"Very, very good for you," he said, "but also very expensive." The Tiki Village is much like any such "village" recreated for the sake of tourism everywhere in the world. We also have them in Canada, with our varied populations and our romantic vision of the past. In Alberta, for instance, there are the Ukrainian Cultural Heritage Village, near Edmonton, and the Calgary Stampede Indian Village, the latter advertised as providing "a great way for visitors to experience First Nations tradition and culture first hand." The messages are always ambiguous, because we know very well that neither Ukrainians nor First Nations members live in the manner depicted in these villages. Yet, some of the tourists' online comments indicated they thought what they saw at the Tiki Village was an actual representation of Tahitian everyday living.

The Tahitian *Office de développement du tourisme* was established in 1966 and has since done its utmost to promote all the cultural activities that would encourage tourists to rediscover the familiar myth, but we will see later how government policies can be at odds with the people's own intentions to display artifacts and provide information.

The attack on the World Trade Centre in 2001 dealt a severe blow to tourism everywhere, as has the last economic downturn. Several big hotel chains had to close down in the last few years, notably the Tahiti Hilton in 2010, leaving thousands of people unemployed. Figures for passenger traffic through the Faa'a Airport in Papeete (which also include local traffic among the islands) show a steady average of 1.4 to 1.5 million passengers annually between 2000 and 2007. Afterwards, there was a noticeable decline to 1.1 million in 2011.

On the other hand, cruise ships come in greater numbers and much effort is spent to make some islands into ports of call not to be missed. Wherever they arrive, ad hoc markets are created, and craftsmen travel from neighbouring villages or even other islands to sell their shell or seed jewelry, fabrics, quilts, pareus, and carvings. The quality is usually good, and prices are assessed accordingly; bartering is not encouraged, which speaks well for the artisans' pride but not for their sales acumen since most also require cash payment, when tourists might expect being able to charge their purchases. This source of revenue accounts for a significant portion of the local economy.

So far, nothing seems to distinguish Tahiti from other islands with exquisite natural surroundings, so it is important to single out any unique characteristic it may have, lest the sun and the diving in Tahiti blend in travellers' mind with the sun and the diving in any other "island paradise." Naturally, we know the reputation of the island and her unique and distinctive feature. The Tahitian message has always been the same. "Tahiti, where love lives," is a popular slogan adopted by the Tahiti Tourist office to promote the islands to the American market. Even when not formally stated, the licentious motto is implied. In 1981, in the course of an East-West Connection television program from Los Angeles, Tahiti's Minister of Tourism was asked, "What is the racial breakdown of the majority of the people?" He replied, "We used to say in Tahiti that the whole world slept with Tahiti" (Kirk, 2011:584). Given the island's history, the statement is probably true, since from navigators to whalers to traders to settlers to artists to young French recruits to tourists, who has not tasted the joys of an island idyll while visiting Tahiti? The modern *demis* are proof enough that the tradition continues.

Kite Makers vs. Directors of Tourism Development

The only recourse for Tahiti to outbid potential rivals for the tourists' American dollars has been to build on her past and the images that served to seduce the navigators: dances, tattoos, *vahines*, etc. In rediscovering those former cultural characteristics once destroyed, a new pride was awakened and a powerful desire to reintegrate them into everyday life. Sometimes convenience prevailed and modern life continued to provide the necessary resources of technology, but at the cultural level, the old ways served to enhance and strengthen the Tahitian spirit by becoming once more a valid part of their lives—even if was often difficult to preserve the authenticity of these ancestral practices.

There have been conflicts between grass roots efforts to revive an interest in some aspects in Tahitian culture and the territorial government's own attempt at taking over all aspects of tourism, paternalistically certain that it knows better what appeals to tourists. It may be right, since awakening the tourists' interest was neither the initial nor the more important goal of the local residents. An example given by Kahn (2011:155-80) is particularly relevant since the site in question is one of much touristic interest—an interest no longer guided by the cliché evolved from the myth of "the island of love," but by a genuine curiosity about Tahitian culture. It concerns the Fare Pote'e and the nearby *marae* in Maeva, on Huahine, an island most tourists find different from the others because it appears more genuinely Polynesian to them.

The Fare Pote'e was originally a meeting place for the village. Rejected by the missionaries, abandoned by the villagers, and replaced by a church, it was later relocated, rebuilt by local residents, and transformed by them into an active cultural centre in 1972. All this was accomplished under the careful guidance of Dr. Sinoto, an archaeologist from the Bishop Museum in Honolulu. The musician and poet Bobby Holcomb, on the island since 1976, had also been involved with making a success of Fare Pote'e as a cultural community centre. He inspired elders to come and demonstrate their crafts, teaching children in Tahitian (rather than French) about their culture. Holcomb remained an active supporter of the centre until his death in 1991.

Chantal Spitz was part a group interested in 1993 in restoring Fare Pote'e, by then in poor condition, and turning it into a cultural heritage centre, with a small museum, craft demonstrations, and dance performances, also serving as a cultural interpretive centre for the nearby *marae*. This movement, known as *Opu Nui*, intended to rebuild Fare Pote'e and develop tourism at the grass root level, with the people from Mavea controlling the preservation of their heritage. When the time came for *Opu Nui* to seek government funding, an official delegation came to visit the site and complained that it was not very "attractive." Whereas *Opu Nui* wanted a place where local people could experience various aspects of their own culture and work in the museum and where visitors could also learn about Maeva and Huahine, the government representatives only appeared interested in providing a site that would attract tourists. There were further discussions and conflicts about a possible relocation, as well as suitable events to celebrate the reopening of the centre. One of the bones of contention was the government's desire to see something more "spectacular" than what *Opu Nui* had in mind, such as "the re-enactments of religious ceremonies on the *marae*." However, restraint won and, after it reopened, the centre was successful with the residents as well as the tourists who seemed unanimous in their appreciation of the Fare Pote'e facilities.

The centre was destroyed by a cyclone in 1998, only eighteen months after its inauguration. While the villagers were coping with the aftermath of the cyclone and still expecting to rebuild Fare Pete'e one more time, the centre was moved to what was deemed to be a more appropriate site and rebuilt as a "scientific information centre" for tourists by the government intent on promoting tourism in Huahine. Dorothy Levy, an early member of *Opu Nui*, commented that "the Fare Pote'e, which was originally meant to be something emotional for the soul to enjoy, has been turned into something intellectual for tourists to digest." It was not perhaps unusual for a successful village initiative to be taken over by the government, but it had deep emotional repercussions.

The restoration of the *marae* followed a nearly similar path. The Bishop Museum archaeologists had originally conducted some investigations on the *marae* on Huahine. Dr. Sinoto, in particular, discovered some two hundred *marae* on the island, forty of them near Maeva, and also succeeded in demystifying the sites for many villagers who had been fearful of approaching them. He also enlisted their efforts in restoring them in a suitable manner, in accordance with their ancestral spiritual meaning.

Later, when it became apparent that tourists visiting Fare Pote'e also enjoyed visiting the nearby *marae*, the government took over their restoration, rebuilding them to make them more approachable to tourists (raising the ground level, reorienting the site, rearranging the stones), in deep contrast with the respectful investigations of the Bishop Museum's archaeologists of the past. Some panels were also erected, bearing descriptions and interpretations of the *marae* in Tahitian, English, and French. They are clearly expressed and provide a superficial understanding of the island's culture. The *marae* had in fact become outdoor museum exhibits, with the spirit drained out of them.

Descriptive Panels at Taputapuatea Marae (Raiatea)

Since the original artifacts had been destroyed at the instigation of the missionaries, the French craftsmen in charge of restoring the sites and explaining the elements of the earlier rituals often had to use sketches made by eighteenth-century artists to reconstitute them. Dorothy Levy's comment about the Fare Pote'e is equally valid for the "restored" *marae*. Devoid of emotion for the soul, they are simply intended for the tourists' intellectual curiosity. Kahn opposes those who initiated and participated in the original project ("villagers, fishermen, dancers, artisans, kite makers, and others") to those who took over ("local politicians, government presidents, ministers of culture, high commissioners, and directors of tourism development").

Many in Tahiti are involved in handling and sponsoring touristic endeavours, from the government, to the Université de la Polynésie française that promotes cultural events, to genuine grassroots efforts to revive now obsolete customs and some festivals. Altogether, there is a strong Tahitian affirmation in the general Polynesian cultural revival, with many active cultural associations. The first, la Maison des Jeunes-Maison de la Culture de Papeete was created in 1970 to encourage cultural and artistic creation in Tahiti. Five years later, l'Académie tahitienne, Te Fare Vana'a, was established to promote the study of language. In 1980, two new institutions were created: le Centre Polynésien des Sciences Humaines and l'Office Territorial de l'Action Culturelle (Te Fare Tahiti Nui) replacing the one created in 1970, and le Centre des Métiers d'Art, designed to research, conserve, and teach the ancestral techniques to young people, created in 1981.

Tahiti also participates in cultural and artistic activities jointly with her neighbours, and underlying political sensitivities are no doubt at work in this Ma'ohi revival. Among the initiatives figure the Festival of the Pacific Arts, started in May 1972 and hosted by various Polynesian countries. It was Tahiti's turn in 1980, when the delegates came to Papeete for the fourth meeting. In 2006, New Caledonia and French Polynesia were admitted as associate members of the Pacific Islands Forum. In September 2007, members of Polynesian royal dynasties held a three-day meeting at Taputapuatea *marae*, the most sacred spot to ancient Polynesians on the island of Raiatea. Former monarchs and chiefs came from New Zealand, the Cooks Islands, Wallis and Futuna Islands, Rapa Nui (Easter Island), New

Caledonia, and Hawaii. All present issued a declaration that they would work to preserve Polynesian culture.

This effort is illustrated through the most popular event in Tahiti, the *Heiva,* lasting most of July and intended to celebrate ma'ohi culture. It draws large crowds who enjoy watching or participating in craft exhibitions, games of skill and strength, singing and dancing competitions, canoe racing, and spear throwing events.

Traditions

Preserving a culture naturally means preserving or restoring its traditions. In the case of a culture stymied by traumatic events, some of these traditions may have been eradicated. Often, tourism requires them to be reintroduced to appeal to the exotic dimension visitors seek while travelling. Some traditions are so deeply rooted that they have survived or have no difficulty being revived, while others have been irremediably lost. Those that have been recreated have the double purpose of enhancing cultural pride while interesting tourists, and we can review in this context the place currently occupied by once intrinsic elements of Tahitian culture such as kites, outrigger canoes, tattoos, dancing, and food.

Kite-flying

The tradition of making and flying kites was officially re-initiated by the islanders for one specific occasion, the opening of the Fare Pote'e in Huanine in September 1996. The old *Ma'ohi* tradition, abandoned for many years, needed some research before being reintroduced. The event was so unusual that an oral historian from the *Musée de Tahiti et Ses Iles* in Papeete came to Huanine to witness the display of the large kites, often shaped like sea turtles or manta rays. One contest was for the children, who had made their kites in school, the other and highly competitive one for the adults from some thirty villages.

Whether this rediscovered pastime is once more part of the islanders' life is unclear. However, it must have remained an occasional tourist attraction, if one relies on an exchange on the *Tahiti Travel Forum* advertising kite-flying on Maeva in December 17-18, 2011. Someone asked whether

Dorothy Levy would be there and was reassured that she would be, adding, "Dorothy is the one that researched it a few years ago and found the very old men who knew how to do it. They had a huge contest back then. This one has less people."

It is difficult to understand how the custom could have fallen in such oblivion, when we know its profound past significance. An Anglican visitor described the ritual in 1888. "In times of peace, kite-flying was a favourite amusement of the Hervey Islanders. Kites were egg, club, or bird-shaped. The first flown was sacred to the gods. Each kite bore a name and heraldic device. Only chiefs and aged men might indulge in this pastime. Tears of joy would be shed by the victor, to whom was given the largest share of the grand feast that followed. During the flight of the kite (or 'bird'), it was customary to chant a song, composed for the occasion, which was supposed to aid its ascent" (Religious Tract Society, 1888: 397).

Many younger people show a strong interest in reviving the skills of the past. However, it is difficult to assess to what extent their interest is a sign of political involvement, as it would have been for the Tahitians of the 1960s (during the emergence of the cultural revival as a stance against the nuclear politics of the French) or in the 1980s (when economic conditions created the need to foster tourism and display traditional lore and crafts). Rather, these young people may simply relish the past—such as it can be said to be known—for its own sake. They cultivate *reo ma'ohi*, tattooing, stone sculpture and wood carving, ma'ohi sports and games, genealogical recitations, and so on. We may suppose, with Brami-Celentano (2000), that the renewed interest of today's young people is more associated with the knowledge once held by the priests, the traditional chiefs, the *'ariori*, or the *haere po*, rather than by the vast majority of ordinary Tahitians, who often stood as mere victims in an often violent society.

Canoes

Some aspects of Polynesian culture have survived well, albeit rather artificially, such as the canoes. We know the islanders' recognized expertise in outrigger canoe construction and navigation, gathering praise and admiration from the eighteenth-century navigators. Commerson left sketches and descriptions: the smaller ones, from seven to twelve metres (*va'a motu*), the larger, double-hulled canoes used for long-distance voyages, and the war canoes (*pahi tamai*). We know the part they played in Tahitian life was a defining element of their culture. Whether used for long-range expeditions, for warfare, for the colourful display of the *aiorii's* passage from island to island, or as the sole means of transportation, they were a part of the islanders' daily life.

The epic voyages were reported in oral history and are still remembered in traditional poetry, but Louise Peltzer (1993) describes a simpler event: a little girl's version of the preparations for a trip from Tahiti to Huahine, a return to *fenua*. It is the men's job to prepare the canoe and the women's to look after the provisions for the journey. The grandfather, who took the decision to return to Huahine, will also decide on the day of their leaving. He must rely on the winds to help on the journey. The *hupe*, coming from the mountain will carry them to the lagoon, the *maoa'e* will then take over and escort them home. When the day arrives, the large canoe is rolled to the shore on coconut tree trunks and then launched. The men had repaired the worn wood, retied the bindings, and checked everything. New parts have been made for the voyage, including the pandanus sails, the platform that joins the two shells, and the roof of the house that sits on it, with its tapa curtain. The men, after loading the bundles, the bamboo baskets filled with *popoi* and *taioro*, the bunches of coconuts, and all that is required for the journey, make their offerings and their prayers on the marae. All are then ready to leave.

If such a scene were re-enacted today, it would probably be as a tourist attraction. No one travels in this manner anymore. Yet, the mystique of the outrigger canoe is deeply implanted in the Tahitian soul. For many, the voyages of the past have acquired an almost mythical character, mostly because they seem so implausible. How could such canoes, made with tools barely evolved from prehistory, touch upon so many islands in the vastness of the largest ocean on earth? Particularly since the rowers who found their way through the waves had no scientific navigational knowledge and relied only on their almost visceral bond with nature.

Today, nowhere is this deep connection between islanders and their canoes more evident than in the *Hawai'i Nui Va'a,* the annual outrigger race that opposes teams from many nations. It is a grueling event, taking place over three days in the early fall. It is described as the greatest and the toughest canoe race in the world.

The route is always the same and the race takes three days. First is the Hauhine-Taiatea segment of forty-five kilometres, taking about four hours; second, a "sprint" of about ninety minutes over twenty-six kilometres, taking place in the Raiatea-Tahaa lagoon; finally, the four hours from Tahaa to Matina Beach in Bora Bora, a distance of fifty-eight kilometres. The rowers' mood is emotional and they often pray before each day's race. A five-time participant said, "It's the dream of every young man to win that race. It's like winning a war: you enter History!"

Tahiti enters several canoes, some privately owned and rowed by volunteers who practice when they can. But it is a sign of the symbolic importance of winning the race that there are also two "professional" teams, in which the men are employed and train regularly for the honour of competing (and hopefully winning) the *Va'a* for their companies—e.g. Esso and the Post Office, the two great contenders in the 2007 race for which I have records. These teams are expected to win, but when an ordinary team, composed of working men with limited means, arrives among the first three or four, the honour is sweetly felt by all Tahitians. As a rule, they easily beat the other teams, and their success serves to confirm their physical strength, their endurance, and their profound reconnection to their seafaring past (Denjean, 2008).

An initiative from Hawaii has certainly confirmed the Tahitians' expertise at sea. In the 1970s, a number of Hawaiians founded the "Polynesian Voyaging Society" with the goal of crossing over to Tahiti on a traditionally built vessel, using the ancient art of celestial navigation, now lost. Herb Kowainui Kane, one of the co-founders, explained the basis of their quest. "No culture can subsist without its objects. When an object is lost, then the memory of it becomes vague, then distorted, and eventually becomes just a fable" (Anthony, 2010). As a first step toward recovering the lost art, they found on the atoll of Satawai in Micronesia a master navigator by the name of Pau Piailug. He knew the old Polynesian ways and agreed to go with their group of novices and teach them to navigate without compass, instruments, or maps.

On May 1st 1976, their Polynesian-style seagoing vessel, the *Hokule'a* sailed from Hawaii on its way to Tahiti, where it arrived thirty-one days later, greeted by seventeen thousand exultant Tahitians. Her crew had maintained their course solely through observing the stars, cloud formations, sea bird flights, and ocean swells. This voyage had not been without problems, causing Pau Piailug to withdraw from the project and return to Micronesia. As one of the acknowledged six master navigators left on earth to use celestial navigation, his contribution to the Polynesian Voyaging Society was indispensable and he once more agreed to go to Hawaii and continue with his teaching, because "it's important that men know the things of the old days." In 1980, the *Hokule'a* set sail again for Tahiti, with a Hawaiian navigator at the helm. The voyage was a success and sailing continued through the Polynesian Triangle. In 1985, *Hokule'a* once more sailed to Tahiti, and continued on to Samoa, Tonga, Rarotonga, Aoteroa. In 1992, it even reached Rapa Nui, the most isolated island on the planet. They even went further. The *Hokule'a* was accompanied on a her 1995 voyage by two other voyaging canoes, the *Makali'i* and the *Hawai'iloa;* the latter travelled north along the American coast to visit the Alaskan villages that had provided the large trees used to build the canoes.

Hokule'a and the outriggers that accompanied her or even sailed on their own *(Alingano Maisu, Makali'i, Hawai'iloa)* had amply demonstrated how early Polynesian navigators were able to discovering and colonizing the far-flung islands within the Polynesian Triangle. The legend of Hawai'iloa, the founder of Hawaii and Maui, after whom one of the vessels was named, told how he returned twice to Tahiti, from where he had sailed, to fetch his family and gather other people, particularly craftsmen, in order to help populate the new islands he had discovered. These, he named after himself (Hawai'i) and some of his children (Maui, O'ahu, Kaua'i), as he also named the stars that had accompanied him after his navigators and steersmen (Kamakau,1932). The narrow ties between Tahiti and Hawaii were thus established as well as the ability to find their way back and forth between the two archipelagos, variously guided by the red star hoku'ula (Aldebaran) and ka hoku loa (the Morning Star).

Nowadays, canoes have mostly lost their utilitarian function, to be replaced by more convenient engine-driven crafts. However, this does not mean that Tahitians have lost their seafaring skills. Danielsson, while sailing in 1949 on a schooner from Papeete to the Tuamotus, relates a joke then popular in Tahiti. A captain says to the helmsman, "You see that cloud on the horizon? Steer on that." Then he goes into his cabin and lies down for several hours. When he comes back on deck, he says to the helmsman, "The birds are flying south-south-east. That means that there's land somewhere there. Follow them as well as you can." He then goes down for another nap. As expected, land appears on the horizon. When the skipper comes back, he says, "Let me see, there's an entrance with two rocks in the middle. That must be Tahanea." Or he says, "No palms on the south side. That must be Toau." But, if he does not recognize the island, he simply goes ashore and asks where he is.

At the time, charts were often so poorly drawn that the skippers did not have much faith in them and, as a result, did not use sextants, logs, chronometers, and handbooks as much as they could have done. Like their ancestors, they relied on experience and traditional knowledge. One skipper explained, "It's a simple matter to hear from the roar of the surf how far one is from land. When the noise is equally loud from both sides naturally I am in the middle of the straight, and if one listens properly there's no difficulty in keeping a course."

All Polynesians are first-class seamen, affirms Danielsson. He explains that Tuamotu skippers never took a crew with them when they went to San Francisco to fetch or buy boats. All they did was look into all the doss-houses and shelters, and grabbed whatever Polynesian they could find there. Whether seamen or involved in other trades, all were born sailors and would perform equally well at sea. One exception to this unanimity of skills was when it came to hiring an engineer, a trade in which most Polynesians had little interest. One skipper found a solution: he hired a cook, assuming that at least he would be able to bear the heat in the engine room even if his mechanical skills were non-existent. Apparently, it worked reasonably well (1954:9-11).

This is all very well, and a light-hearted banter reported by Danielsson. Certainly the mechanical ignorance he mentions does not relate to contemporary experience (three generations have gone by since he wrote). However, it all speaks to perception rather than reality and, in the case of Tahiti and more generally, Polynesia, imagination and interpretation, often disguised as facts, count for much—whether we think of the original Arcadian myth or the language of modern tourism. In the present case, the deep relationship with the sea, whether inherent to myths, flaunted in touristic displays, or naturally incorporated into ordinarily life, is an integral part of the Polynesian spirit.

Tattoos

One of the most famous illustration of Marquesan tattooing is reproduced in A. J. von Krusenstein's *Voyage Around the World 1803-1806*. The man carries over his right shoulder a heavy mass, and a gourd-like container hangs from his left hand. Apart from his elaborate headpiece and neck plate, and the thin cloth draped over his hips and hiding his penis, he is entirely naked. The whole body is covered with intricate designs, both geometrical and figurative, and the facial features are outlined. The body is so covered that the ink, from the soot of the candlenut, darkens the skin to the point of making it appear almost black. Another intricate example from Nukahiva" is found in G. H. von Langsdorff's *Voyages and Travels in Various Parts of the World* (1813).

Other contemporary sketches indicate that these men's tattoos were not the exception but seemed to have been the rule for highborn men. W. D. Hambly reproduces in his *History of Tattooing* whole-body patterns from the Marquesas, in which the body actually seems to disappear and only serves as a canvas for the intricate artwork.

Tattoos from the Marquesas (Early 1800s)

Wallis, Bougainville, and Cook all remarked on the tattoos they saw in Tahiti, particularly on the deep blue or black paint or dye with which both men and women decorated their thighs and buttocks. Indeed, when Omai went to England, the English were most taken with his physical ornamentation.

Upon seeing Tahitian tattoos for the first time, Cook and Forster commented that the coverage was so thorough that it was "like a coat of mail" and that the patterns were inked so close together that they hid the elegance of the wearer's body. No body part seems to have been forgotten. Some of the men were tattooed on the scalp, under their hair, and even on the gums, tongue, and the head of the penis under the foreskin. Forster observed that the islanders favoured geometric motifs, consisting of a variety of "blotches, spirals, bars, checkers, and lines." Like their wood carvings, the tattoos were often divided into sections; although each section could be different, the overall effect was one of balance. Generally speaking, there seems to have been some intent to cover most of the body, but Sparrman observed an unusual tattoo on a young man that started in his hair on the right side of his head, wound in one unbroken line across his forehead, nose, mouth, chin, throat, chest, and stomach, and ended up on the big toe of his left foot (Cameron, 1964:167–68).

Charles-Félix-Pierre Fresche (2007), who sailed with Bougainville on the *Boudeuse,* was most impressed with the tools and technique employed. He wrote in his journal that the instrument used looked like a comb and was made of very small pieces of sharpened shell attached to the end of small stick, about half a foot long. The "comb' is dipped into a dye and applied to the skin by use of a mallet. The indelible dye then penetrates the skin, which remains sore and swollen for a few days.

However intricate some of these tattoos appear to be, there was very likely little left to luck or artistic freedom. The patterns are said to be intrinsic to the cultures of the Pacific and their motifs are traditional. The tattoos served to place the individual in a particular community and often represented his all-important genealogy, his most admired achievements, and possibly conveyed ideas of the sacred, since they were often applied according to strict rules by specific people, the *tahua.*

After being banned by the missionaries, tattooing was reintroduced in the 1980s, with a preference for the traditional method using the same bone or tortoiseshell needles mounted on a wooden handle, dipped in charcoal ink (usually burnt candlenut dissolved in either water or oil), and hammered in. It was banned almost immediately, owing to the difficulty of sterilizing the instruments, and was replaced with modern tattooing methods. The traditional method is reputed to have been very painful, and Joseph Banks commented that when Polynesians showed off their tattoos, he did not known whether it was to display the artwork or to prove their ability to endure the pain of obtaining it.[27]

Not all Europeans admired tattoos. Melville was relieved that women were spared from the manifestations of the "barbarous art," inflicted so remorseless "upon the brawny limbs of the warriors of the tribe." Among them was Fayaway, whom I could have included as one of the women who served to nurture the Tahitian myth. Melville was definitely in awe of the young woman's charms and relieved that tattoos had not defaced her beauty. "All the tattooing that the nymph in question exhibited upon her person may be easily described. Three minute dots, no bigger than pin-heads, decorated either lip, and at a little distance were not at all discernible. Just upon the fall of the shoulder were drawn two parallel lines half an inch apart, and perhaps three inches in length, the interval being filled with deli-cate executed figures." These narrow shoulder tattoos always reminded him of officers' "stripes of gold lace worn... in lieu of epaulettes to denote their rank" (1982:107). White men may not always have appreciated women's tattoos, but native women felt differently. Robert Louis Stevenson wrote about a white man who fell in love with a native woman, "When solicited, she said she would never marry a man who was not tattooed: we looked too naked!"

The missionaries abhorred tattoos, as they did any other trace of the importance of the older Tahitian culture. But were they successful in banning them entirely? Dr. Frederick Bennett, who sailed the world on a whaling voyage (1833-1836), illustrated his *Narrative* with many sketches and was able to reproduce tattoo patterns from the skin of a Tahitian and a Marquesan. Naturally, banning tattoos would only apply to the future, for little could be done about the ornamented men still walking about. Unlike the ephemeral dancing, tattoos were there to stay. Even if new tattoos could not be openly drawn, they remained in evidence until death, a living testi-mony of a suppressed culture.

In "Tattoo," Rai a Mai relates how a modern child is told about tattoos. She had seen an old Marquesan woman adorned with them and believed they had enabled her to communicate with the elements. Impressed, the child wishes to have tattoos so she, too, could understand the wind, the clouds, and the sea, and "to hear what others could not hear." Her father said, "We used to tell our story on our body. And people and heavens would know who we were. But nowadays, stories and words are written in books. The words are caught directly from our memories and written with ink on paper. You don't need to catch the words in the clouds from Heaven anymore. They are here!" And he pointed to her forehead. Her grandmother said, "Today, no one has Polynesian tattoo on their body anymore." She admits that some have tattoos brought back from the army but, rather than speaking of war, they often only speak of women, love, and broken hearts (a heart pierced with an arrow); men also bring back tattoos from prison that tell of jail stories. She sadly concluded, "They are unfinished designs. In fact, nobody knows how to tattoo the way our ancestors did. They have forgotten" (Stewart, Mateata-Allain, Mawyer, 2006:181-82).

The little girl grew up and started seeing more and more tattoos around her, on both Tahitians and white men. Protestant missionaries had ended the practice, but in the 1980s, during a cultural renaissance in the islands, there was an attempt to revive it. Almost everyone then had a tattoo, but for Tahitians the revival took on a different meaning, "Every Polynesian wanted, stamped into the skin, a sign of cultural belonging." In *Breadfruit,* Vaite describes the tattoos worn by members of Materena's family and friends. Her cousin Moori's tattoo shows a red and green dragon spitting fire. The comments he receives are not particularly flattering: such tattoos make you look as if you have been to prison or are Chinese. Another cousin, James, has a heart pierced with an arrow. Pito, Materena's husband, whose totem is the gecko (he is actually afraid of them) has one put on his thigh, to show "his identity." Her own godfather, Hotu, had tattooed on his right shoulder the name of the woman he loves, Imelda. None of these tattoos seem connected to anything other than the ordinary circumstances of life, except for the gecko tattooed on Pito's thigh. Actually, given the current fad for tattoos, the same patterns are probably drawn everywhere in the world. However, it is true that the most popular tattoo motifs currently etched in Tahiti are culturally relevant, such as the turtle, gecko, ray, shark, and dolphin.

Most Tahitians seem to have a tattoo—certainly nothing like the extended version of their ancestors, but definitely a tattoo along the arm or the leg, of any other visible part of their bodies. Naturally, for people working in the tourism industry (dancers and drummers, hotel and restaurant employees, etc.) tattoos are almost a requirement, since the tourists expect to see tattooed Polynesians.

A young man told me that in the olden days "your tattoo was almost like an identity card; it said everything about you. Now, your tattoos have no special meaning, it's just a desire to be identified as Polynesians." On the other hand, on the documentaries on the Marquesas cited here, most men were shown to be beautifully and extensively tattooed. Since they were engaged in traditional arts (a tiki sculptor, a healer, a drum maker), it is possible that they were more deeply connected to the significance of tattoos in their culture.

Facial tattooing takes another form in this oddity reported by Gauguin while attending a wedding. He saw there a hundred-year old woman with, on her cheek, a tattoo "suggesting the style of a Latin letter." Intrigued, he enquired, because he saw in it a European hand. "'Formerly,' they told me, 'the missionaries, zealous against the sin of the flesh, marked 'certain women' with a seal of infamy, 'signet of hell.' It covered them with shame, not because of the sin committed, but because of the ridicule and the opprobrium associated with such a 'mark of distinction'" (1994:98). I have not read any other mention of this form of tattoo by missionaries. It seems odd, to say the least, but perhaps not entirely out of keeping with the zeal and severity they sometimes showed.

Dances

Once banned as lascivious exhibitions, Tahitian dances have been resurrected as a strong drawing force in modern tourism, backed by a government that relies on this income to meet endemic budgetary shortfalls. It would appear as if, miraculously, such dances were no longer reprehensible. In fact, they constitute now the essential attraction of every government-sponsored touristic event.

Of the dances in Tahiti, the traditional *upa upa* was deemed the most indecent, as it was performed by young men and women mimicking the motions of sex. New versions exist now that dancing has been reinstated, both as a popular pastime and as a folkloric performance. Some dances have been lost and new dances have also evolved from the more traditional ones. Most are accompanied by percussion instruments like the *to'ere* (a slit log which is struck by sticks), the *pahu* (a large vertical drum covered with shark skin) and the smaller *fa'atete* drum.

Two types of dancing seem to co-exist in Tahiti today, only connected through illusion. One is part of everyday life. Tahitians dance because they have always done so, both in the free days of their past and much less openly after the missionaries' ban. Many of the dances may have lost their spiritual meaning, and it is also sadly true that many of the old dances have simply been lost. According to Jacquie, a Marquesan looking for lost dances in order to "recover the taste of the culture," modern Marquesans could only reconstitute ten of them, although the missionaries had reported the existence of seventy distinct dances, some with very important significance (Lefèvre, 2011).

Nevertheless, Tahitians still dance for pleasure and among themselves, and sometimes compete against each other. During the Heiva festival, held every summer in Papeete, people from several islands meet to dance and compete.

The second type of dancing is the one performed in tourist shows. Designing, choreographing, and promoting these shows is big business in Tahiti. A friend of John Stember's (one of the main photographers of postcard *vahines*), Tumata Robinson, is one those involved in it. Half Tahitian, half Russian (the father, naturally), she is a Papeete choreographer and designer who organizes sumptuous shows displaying local dancers and promoting the image of the lovely Tahitian woman, totally free of body and spirit.

Another important player in the Tahitian tourism industry is Paulette Vienot, a French woman, who was instrumental in using traditional dancing as a focus of tourism by creating Tahiti Nui, the first official dance group that toured the United States for several months. What was promoted, even if based on traditional Tahitian dancing, was a form of dancing specifically designed to appeal to tourists and fulfill their expectations.

Marietta Tefaataumara, once the leader of a Huahine dance group, explained to Miriam Kahn the difference between the manner Tahitians dance when among themselves and when their dancing becomes part of a remunerated hotel show. What happened, she said, was that "The songs and dances began to lose their meaning. Normally, music, words, and special hand gestures go with every song. Tahitians know these details because they live inside the dance. It's their culture." On the other hand, when the constraint comes to produce a show for tourists and the meaning has no relevance for the watchers, the dancers no longer behave as they normally would. They borrow various elements "from everywhere" to produce something that will appeal to the spectators. "The girls hike up their *pareus*. They smile and show their teeth. They do some hand gestures. Tourists like that because it's an attractive show to watch. It's pleasing to the eye. It makes a good video. But the meaning is gone" (Kahn, 2011:118). Indeed, it would be impossible for such dances as the graceful *aparima,* a story-telling dance, to transmit its tale to people ignorant of the language and unaccustomed to reading the meaning of traditional gestures.

The commercialization is pronounced and hard to miss. The same Marietta Tefaataumarama can barely hide her contempt when she speaks about Les Grands Ballets (a government-sponsored group that travels overseas to promote tourism to Tahiti) who perform with an Air Tahiti Nui banner behind them on the stage.

For many tourists, the experience of watching a dance and drums performance on stage or in a hotel has become an integral part of their visit to Tahiti. It takes some naivety or good will on the part of tourists to mistake this dancing for the real thing, but some probably believe it is a valid introduction to the culture. Kahn comments that watching these staged performances is as comprehensive an insight into the essence of a culture as only showing Frenchmen eating snails or Dutchmen walking about in their clogs.

On the other hand, many probably know that it is with dancing as with every other recovered cultural re-enactment intended to whet their appetite: nothing is really what it seems once it has been translated into a tourist performance. There appears to exist a connivance of sorts between the tourists, the promoters, and the Tahitian performers. The promoters propose a somewhat fictitious exhibit, to which the islanders agree because it does not really affect them. If dancing in hotels is good for tourism and the tourists have a good time, why not do it? It is a source of income and so little of the culture is actually represented that it is simply a show without much meaning. The tourists usually enjoy the performance, even if they realize that what they see is only designed to satisfy their own expectations. They appreciate the shows for what they are: a display of graceful feminine movements or a session of virile drumming. Nothing more and nothing less. For instance, one of the charming dances I watched was performed by a woman gracefully showing the various ways of tying and wearing the *pareu*. It was a clever combination of two different traditions, exclusively for the benefit of outsiders on whom the cultural relevance of dancing and the significance of wearing the pareus were otherwise totally lost. We just enjoyed the show.

It is likely that most Tahitians are rather indifferent to the tourists' responses to these shows. These are purely commercial exercises designed to bring in additional income. It is mostly accidental that they also presume to speak of the culture. But not everyone looks upon these performances with equanimity or indifference. As one would expect from Chantal Spitz, determined as she is to preserve the character of her culture and, to the extent that it is still possible to do so, its purity, she has something to say about these shows that pretend to project an image of Tahiti to suit the stereotype familiar to the audience. "What is the worst for me, is that we have completely integrated the myth and we have arrived to the ultimate stage of colonization: we repeat what is said about us and we behave so as to *please* others. And it's worth seeing how our country is perceived abroad. We seem unable to send writers, and all we can send is dancers, and half-naked tattooed guys, who shout on the stage, who will grunt like wild pigs. It really bothers me" (Pernoud, 2012).

Food

Like any other cultural manisfestation, food is anchored in tradition. This is most evident in food taboos and prevalent cooking customs, their origins sometimes forgotten.

Clariza Lucas writes, "There are three ways to destroy a people's identity: by fighting against them with weapons, by substituting their language, and by changing radically their nutritional habits (1992:2)." The first two are certainly deliberate attempts at changing a culture, "destroying" it as the author insists, but what of the third, the changes in food cultivation and consumption? Must the change in nutrition necessarily been deemed to have a Machiavellian source? She adds, "France has become a real expert in these three methods," thus confirming that, in her mind, all three actions contribute equally to the destruction of one particular culture and that this destruction is willfully brought about by another, dominant, culture.

It is indeed well known that a nation's age-old nutritional habits can succumb very quickly to the importation of new and more convenient foodstuff. In 1995, I was briefly associated with "The Pacific Islands Community Nutrition Training Project," a joint project of British Columbia's Simon Fraser University, the University of the South Pacific, and the South Pacific Commission. Twelve nations were involved in the project: the Cook Islands, Fiji, Kiribati, Marshall Islands, Nauru, Nieu, Solomon Islands, Tokelau, Tonga, Tuvalu, and Western Samoa, and involved a population of 1.3 million. In spite of the distances, the languages, and many other differences, these Pacific Islands were united in the adoption of the World Health Organization's objectives of "Health for all by the Year 2000" and "Water and sanitation for all by the year 2000."

In 1982, Fiji had noted that twenty-three percent of its children under the age of five were undernourished. The Marshall Islands' 1985 figures for the same group were thirty percent. Among adults, obesity and illnesses related to poor diet were also increasing. Finally, in 1987, a United Nations study put the problem into an international perspective by showing that South Pacific Islanders displayed the world's most rapidly increasing rate of malnutrition.

The departure from traditional diet and the adoption of Western food were identified as the main causes for the deterioration of health through malnutrition. New trends had emerged in food economies, and convenience food supplanted traditional foodstuffs and methods of food preparation. For instance, in Fiji, a land surrounded by waters then teeming with fish, the small amount of fish eaten was mostly canned mackerel; babies were bottle-fed; white flour, sugar, and rice had replaced the far more nutritious taro root and yams.

An earlier 1975 report had already recognized specific concerns about the small island of Nauru, near the Solomon Islands. The report, by the Commonwealth Department of Health (Canberra) was titled "Dietary Survey of Nauru," and found that "the Naruan diet now consists almost entirely of imported foods." Alarmingly, fish was hardly ever consumed anymore, and the Naruans' diet consisted mainly of polished rice, bread, canned meats, canned milk, and sugar, the latter two actually constituting twenty-five percent of their food intake. In contrast, twenty years earlier, a "General Dental Survey of the Island of Nauru" had found the Nauruans to be "healthy, big and strong, with strong teeth and jaws from chewing coconuts, pandanus, and raw fish and bones."

The deterioration of diet in Nauru has been so outstanding that Jared Diamond considers it a special case. He gives the percentage of diabetics in the population as follows: 0 (1952), 41 (2002), and 31 (2010), the latter figure not necessarily reflecting an improvement in their health but, as Diamond suggests, more likely representing the 2002 diabetic patients' deaths that occurred in the meantime. Pacific Islanders seem to have fared worse than any other population group he reviews, showing either an evolution over time or the difference between traditional and urban populations (for instance, with the telling figures of diabetics in Papua New Guinea—traditional: 0 and urban: 37%) (2012:439).

The reason for this extraordinary difference is usually due to economic conditions. In the small island of Nauru, first annexed by Germany in 1888, then occupied by Australia in 1914, and an independent republic since 1968, the traditional diet was based on agriculture and fishing. Although they sometimes faced episodes of starvation due to droughts and poor farming conditions, the inhabitants were on the whole reasonably healthy. In 1906, large deposits of phosphate were discovered under the poor soil of Nauru and the islanders were given appropriate mining royalties. Because of these generous revenues, there was no longer any need to bother with farming and fishing; instead, they became particularly fond of sugary foods, which they consumed in large amounts. Their weight increased. They also became sedentary and used cars for transportation. They are among the world's richest people (their individual royalties had risen to about $23,000 in 1968 when they became an independent nation) and also the Pacific islands' most obese population (Diamond, 2012:436).

How does Tahiti fare in these matters? We remember the abundance and variety of fruits and vegetables the navigators found in the islands, and how much fish could be had from the ocean. The islanders had all they needed for a healthy diet: taro, yams, breadfruit, and all types of seafood. Fruit trees were abundant, and small pigs were bred. In fact, with little labour involved, the Tahitians' diet was extremely rich and varied at the time of contact, as scurvy-ridden sailors found out. Thanks to it, the islanders were able to spend more time developing a rich oral culture.

We have a good idea of what a Tahitian feast consisted. Spitz describes the preparations for the visit to the village of French officials in 1914. After airing in the sun their best Sunday outfits and preparing crowns and garlands to greet the visitors, "of course, the men kill a pig, catch crabs, fish and crayfish, dig up taro, ufi [yam], 'umara [sweet potato], mei'a [banana], fe'i [mountain banana], 'uru [breadfruit] and put it all in the traditional oven, the ahima'a" (2007:25).

Gauguin describes two festive meals he attended, with a similar menu. He and Tehura, unannounced, once visited neighbours. "[A little pig] was slaughtered, and two chickens were added. A magnificent mollusk caught that very morning, taros and bananas, made up the menu of this abundant and tempting repast." Another time, they attended a wedding at Mateiea. "The meal at Tahiti, as I believe elsewhere, was the most important part of the ceremony. On Tahiti, at any rate, the greatest culinary luxury is displayed in these feasts. There are little pigs roasted on hot stones, an unbelievable abundance of fish, bananas and guavas, taros, etc." At the same wedding, was also illustrated the traditional oratory present at all formal occasions. "After this [meal] the speeches began. There are many of these. They are delivered according to a regular order and method, and there is a curious competition of eloquence" (1994:77-78, 81-82).

Vegetable-fed dogs were also deemed to be a delicacy. Cook and Banks tasted it steamed in the earth oven between layers of hot stones and green leaves. "Few were there of us but what allowed that a South Sea dog was next to an English Lamb," wrote Cook (Beaglehole, 1974:185). The taste for it seems to have endured as Danielsson wrote in 1954, "The Raroians still regard dog's flesh as a delicacy and gladly pay a couple of hundred francs for a fat dog" (1954:81).

But the cultivation of food changed. Danielsson writes of a conversation he had in the 1950s with a Marquesan who asked him, "Have you noticed that there are hardly any pigs? Do you know why? No one will feed a pig, for as soon as it's fully fattened, it's stolen." It was the same with the crops. "There's no sense in growing anything if one mayn't keep it for oneself. So all the people eat tinned food and white bread which they buy from the Chinese. Most of them don't even fish. It's too much trouble." On the other hand, a little more effort was exerted for another type of consumption. The same man said, "Do you see that there are steps cut right up to the top of nearly all the palm-trunks? That's to make it easier to climb up and tap the juices from the flowers. You cut off a stem and hang a jar underneath. That jar's emptied once a day. Sixty per cent spirit is made from the juice" (1957:95). The two statements may not be entirely unrelated.

What of today? There are several reasons why the Tahitians' diet has changed. Clariza Lucas explains that "the Maohi live essentially from what they fish or gather. After the atomic experiments, the geographic contaminated fishing zone got much larger near the nuclear experimentation complexes of Hao, Fangataufa and Moruroa. Actually, 90 percent of the food must be now imported" (1982:2). It is certainly true that a legitimate fear of contamination affected the consumption of food. However, there were other factors that precipitated a radical change of lifestyle among the Tahitians.

One was the mass arrival of the French during the 1960s, which totally transformed the way of life of the Tahitians, particularly in Papeete. Until then, they had lived in rural areas, had farmed and fished to feed themselves, and trading surplus with neighbours and family. There was little need for money to survive in such an economy. The arrival of the French meant plentiful jobs for the islanders, and the salaries they earned meant the emergence of a cash economy. No longer having time to grow or gather their food, they used their salaries to purchase it. With the demand grew the supply, and Tahiti would soon have to import most of its food.

However, we must remember that we are referring here to Tahiti and, more particularly, to Papeete. The other islands, much less involved in this great swoop of modernization, very likely maintained much of their customary nutrition. But there, as well, there were changes. For instance, the use of personal gardens has been transformed to make the most use of the land in order to earn a living. A friend of Kahn's on Huahine, talking about the way Tahitians had started selling their land, asked, "Where do the Tahitians live? They have no land. No house. They end up living in a tiny shack." As a consequence, "They don't grow their own food anymore. They buy it in the store. If they want Tahitian food, they buy it in the market. I go to the market every Sunday to sell my Tahitian food. Do you know who buys it? Tahitians." This was in 1995. Six years later, the same woman explained to Kahn how she made the maximum use of her garden. "You plant yams to eat. You plant melons to sell in the market. You plant flowers to sell to the hotel. You plant vanilla to sell to tourists. You also bake cakes for people who need a cake to celebrate a birthday. You cook food to sell in the market" (2011:74-75).

Once favourite traditional foods, such as turtles, have now come under official protection. "Eating turtle is part of our culture," says a resentful resident of Raiatea. The Turtle Centre in Bora Bora has forbidden their consumption—which does not prevent local fishermen from eating them, as they do not believe turtles are endangered. What they mostly resent is that turtles are apparently served in restaurants in Martinique and eaten regularly in the Tuamotous (Moneroy-Dumaine, 2012).

What of the celebrated breadfruit in today's Tahitian diet? Even its name—the same in all three languages of the first Europeans to describe it (breadfruit, *arbre à pain, árbol de pan)*—shows that they recognized in it the same basic life-sustaining properties provided by bread in their own countries. To find out the ordinary fare of an average contemporary family often means looking at books written by women about normal household activities, so I turned to Célestine Vaite's trilogy. Materena and her cousin Tapena discuss this very topic in the eponymous *Breadfruit*. Tapera believes their cousin Rita and her mate Coco do not actually have to eat breadfruit because "they've got money and no kids." Materena immediately defends the staple food and Rita's appropriate respect for it. "Rita, she loves breadfruit." The two women then give their favourite recipes for eating it, in a stew in place of potatoes, fried, or baked. Its main quality is that "it tastes nice and it fills the stomach quick." However, there is little doubt that it is not the food of choice, but the one resorted to when finances are low or when savings are required for a big purchase. For instance, Tapera's family "ate a lot of breadfruit for Rose to play the piano" (2006a:102).

Nevertheless, because "you can always rely on breadfruit," it is important when you buy, or even rent, a house, to make sure there is a breadfruit in the yard.

I was somewhat alarmed to see the many positive references made to corned beef, which they often eat instead of meat because it is cheaper. When Materena sends Tihoti, her husband Pito's friend, a package while he is in France, she thoughtfully includes a tape she made for him to introduce herself and her children, four blocks of coconut-scented soap … and three cans of corned beef. It is common for visitors who drop in to be fed the same. When a cousin from Rangiroa visits Mama Noelene, he is urged to eat, so "he sits at the kitchen table and digs into the corned beef."

Not only is it cheaper than meat, but also it is less trouble to prepare than any other food. Materena works outside all day and looks after Pito and their three children, cleans their house, prepares their food, keeps them tidy and happy. She must also visit her mother often and her other relatives. Moreover, she earns a decent salary and can afford the convenience of processed food.

Traditional dishes continue to be cooked with the same ingredients: pork and chicken (sometimes cooked in the *ahima'a* - the traditional fire pit), fish, shrimps, plantain, taro leaves, breadfruit, banana, and the fruits of the ever present fruit trees in almost every garden. These dishes are often seasoned and flavoured with coconut milk and the wonderful local vanilla, now exported all over the world. Such meals are not unlike those earlier described by Gauguin, and they also form the basis of the sumptuous buffets presented to tourists. Tahitian cuisine exists, at various levels of sophistication and corresponding to various levels of income and the presence of a handy garden.

Urban Tahitians sometimes indulge in less traditional meals, if we believe Kahn's description of the tantalizing aroma emerging from the food-vending *roulottes* on the waterfront: "sizzling steaks, chow mein, pizzas and crepes filled the air." In other words, they sometimes eat what we eat ourselves when we look for an easy way out of cooking our meals. It would be hypocritical to begrudge Tahitians the same luxury because they were once the models we created for a natural and healthy way of life.

We should remember that there is difference between the urban and salaried classes, short of time because of outside work and able to afford the convenience of processed food, and the rural population with fewer opportunities to obtain such food and more time for growing and preparing what nature can provide.

Tattoos. Dances. Canoes. Those are probably what most visiting tourists would describe as the culture of Tahiti. For Ma'ohi, it must no doubt rankle to see their rich culture reduced to those aspects of it. All the more so that neither tattoos, nor dances, nor outrigger racing have any real practical function in modern Tahitian society. Their success in international canoe racing no doubt make Tahitians proud of their seafaring history, but this racing is mostly a display of skill rather than a once vital means of transportation and exploration. Dances have been lost and those shown to tourists are only intended to be pleasing to the eye and no longer tells the stories they were meant to relate. The losses suffered by Ma'ohi culture may be best expressed by Chantal Spitz. "You know that we really were a great people. But as we are now, we have lost the dreams of our Fathers, lured away by the glitter of other, different cultures" (2007:138).

On the other hand, new Ma'ohi accomplishments are proudly celebrated. Seventeen thousand Tahitians came to greet the *Hokule'a* after her first crossing from Hawaii. This was the celebration of the renewal of a tradition, but the pride is also strong when successes are not built on the experience of the past. One only needs to have seen in June 2013 the Tahitian football team's totally unexpected goal against Nigeria's six—Nigeria being the ruling African champions—in their Confederation Cup debut at the Estadio Mineirao in Brazil, to be convinced of the pride felt by all Tahiti. They came in as the underdogs, ranking 138th in the world, representing Oceania for the very first time—an honour previously left to Australia and New Zealand. With only one professional player in a team that included four young men from the same family, they came out the crowd's favourites and were able to greet their own performance with the same ancestral gesture they use to celebrate their victories in canoe races.

[XIII]
THE QUANDARY

Quandary: *1. A state of perplexity. 2. A difficult situation; a practical dilemma.*

"A man can find a life for himself here," had said one of Bligh's sailors, a future mutineer who had attempted to make his dream a reality and who may have ended up on Pitcairn Island, or in the dreadful conditions on board the prison-ship *Pandora*, or drowned at sea. No one even knows whether these words were ever actually pronounced, since they were spoken in *Mutiny on the Bounty* by an actor giving substance to a creation of the scriptwriter's fancy. But the actor worded what some of the English sailors had thought in the eighteenth century, and certainly what other people went on to think later about the magic island of Otaheite, and later of Tahiti.

Indeed, many tried to find a life there. Many also wrote about doing it. The Tahitians themselves, who had until then found on their islands a perfectly satisfactory life, were not asked how they felt about this invasion of hopeful dreamers. They were soon to be decimated, converted, and culturally impoverished. Those who resisted were faced with insurmountable forces stacked against them. Then, most appeared to become resigned to their fate, almost second-class citizens on their native soil. However, when the final insult came, not to themselves, but to their land, their mother earth, in the form of nuclear testing, they reacted with violence and bitterness.

Politics have since taken over, replacing the armed resistance of the early days. And writers in abundance, for a culture only so recently converted to the written word, have consistently pushed for a purely Ma'ohi revival. They are the ones who have provided, over the last few decades, the impetus for recreating a potential version of what Tahiti could become.

Spitz and Vaite's books seem to polarize two positions vis-à-vis the experience of living in contemporary Tahiti. While it is simplistic to reduce Tahiti's options to reflect the vision of these two novelists, the authors must have been shaped by ambient thoughts and sentiments, and must find an echo in some of their homeland readers' feelings.

Spitz sees Tahiti still trapped in the consequences of the myth that helped bring the island to its knees. The author experiences a visceral need to reconnect with the "belly" of her land; for this, she would sacrifice all the so-called progress brought by colonization/civilization, not seeing this as a sacrifice at all but rather as a rebirth through which more is regained than is lost. She believes in the power of the ma'ohi spirit, centred in the land, the culture, and the family.

For Vaite, life in Tahiti and her islands is far from ideal, but it is one where nature is bountiful, where people make do, where parents can see their children get a good education (even if it means having to send them to France to receive it) and, once educated, have reasonably good job prospects; where people get modern medical care; and where social services try to meet some of their needs. Her main character is well aware of the abuses of the past, but her realistic approach enables her to enjoy the small pleasures of her life and, thanks to the close connection with her extended family, she feels that the threads of her life are interwoven into the fabric of her community. It is through this interconnection that she feels her Ma'ohi identity blossom. On the other hand—and the ending of her last book seems a little contrived—she is reunited with her French family: Tom Delors and his youngest daughter Thérèse, whom Materena calls her sister, "because there is no half-sister… In Tahiti siblings are siblings full stop" (2007:252).

Vaite is something of an exception in Tahiti's literary panorama by not speaking overtly against the status quo, perhaps because she lives in Australia, and even achieving a total reconciliation with her French side.

Let us go back to the literature of resentment, a label of which I was at first somewhat uncertain but which I must now accept as valid. Michou Chaze is another of its proponents. Born in Papeete in 1950, of a French father and a Tahitian mother, she is another *demie*, who does not mince her words when considering what being a citizen of French Polynesia means. She must pay taxes, bills for her electricity, her insurance, her rent, her social security, her gasoline, her food, the children's school cafeteria and their treats. In fact, she must now pay for everything that used to be free. "I have to pay for bananas and coconuts that I no longer have the right to pick. I have to buy water to bathe and mono'i to perfume myself, and now I must buy my seductive crown of flowers." Most odious of all, "I have to pay for the death of my parents in order to keep the land of my ancestors. I have to pay off the Japanese who are trying to buy us out. I have to pay for a European passport so that Belgians and Italians are at home here. I have to pay for war. I have to pay for the bomb and then pay for the hospital and the coffin. I have to pay to go see our gods displayed under glass in museums" (1990:80). Her resentment is at once political, social, and nostalgic for times gone by. Everything is put pell-mell into the same bag of resentful exasperation: the good (advantages and convenience of modern life) and the bad (cultural loss, political exigencies).

Both she and Spitz are joined in the same desire to do away with the connection with France, whatever the apparent cost. In an interview from her home on Huahine, Spitz quotes her father: "Don't worry. One day those people will go back home." Then, she continues on her own behalf, saying, "When I was young, I was more utopian. I thought that I would not die French. Today I think I will die French, but that perhaps my grandchildren won't." She further reflects in another interview on the possibilities of such a reversal to an older way of life, closer of the essence of Ma'ohi culture. "We can be self-supporting. [We have] fish, fruit, vegetable. We don't need anything else and I want people to be aware that the country can be independent. Life is simple. One does not have to consume much. A minimum of comfort is required, of course, but I believe that living like that is ideal. In the far away islands, maybe it's not too late. In Tahiti, of course…" (Pernoud, 2012).

This is hardly an economic plan for autonomous survival, but it is certainly a romantic vision of a return to Bougainville's very interpretation of the conditions on Tahiti in the eighteenth century, where life definitely appeared to be as simple and untainted as Spitz imagines it. This is the primary side of the Tahitian myth: no longer that of the *vahine*, but the all-encompassing one about Arcadia, where perfect harmony reigned between man and nature.

Mixed with Spitz's inherited memory is a great deal of nostalgic imagination. In his 2003 memoir, *Living to Tell the Tale,* Gabriel García Marquez remarks that life is not what one lives "but what one remembers and how one remembers in order to recount it." This would be particularly true of the cultural life of a group, remembered only through its oral tradition, and now recounted as truth—the explanatory essence of myth.

Living in Huahine, Spitz conceives of a different life but becomes more realistic when she admits that "in Tahiti, of course…" Tahiti, where it would be impossible to abandon all the trappings of modern life and go back to the old ways. One only needs to see the traffic in Papeete, endless lines of modern, expensive, and heavily taxed cars taking their proud owners to work to doubt the latter's willingness to abandon such proofs of economic success.

Others, on other islands, agree with Spitz' feeling that it would be possible to live simpler lives, much as they were once did. In a 2011 Lefèvre documentary, a copra trader would like to tell young people without a job that "the soil is rich, you can plant, you can even breed animals, and preserve the culture of our island." He picks coconuts and dries only a small amount of copra because all he needs is a little money "to equip the house a little, the kitchen, to buy sugar, coffee, and some books." They have all the plants they need to feed themselves. Once more, we come back to paradise on earth, Bougainville's near-fantasy. It is as though we cannot escape from the myth in either its European or its Polynesian version. In the same documentary, Pierre, a sculptor and drum maker, feels the same, although a little more angrily as he thinks of all the "ancestors' objects" kept in museums all over the world. He says, "It's easy to live. All you have to do is plant. And the sea is there," gesturing towards the blue background.

Naturally, picking fruit off the trees, fishing in the lagoon, and hunting wild pigs do not help govern a people. Nor do they provide the essential services expected in the twenty-first century, even on a more reduced scale. Spitz and Chaze do not attempt to draw a blueprint for what an independent government could be. It is not their job to do so. As writers, they see their function as showing anger, revolt, and spreading the nostalgia for the dispossessed land.

Anger is the phase that comes after the grief and shame for the Ma'ohi who had allowed themselves to be stripped of their powers, a sentiment which tended to be the domain of the male poets. Perhaps men believed they had more to lose than women when they thought they had lost a battle. We also notice that they mostly asked questions, the deep questions that anticipate that the answers will be even worse.

"Ma'ohi, are you a man without parents / Are you a man without a land / Are you a man without a government / You did once govern yourself. What does this mean?" (Turo a Raapoto).

"Have I lived so long to witness this shame? / To see perish the essence that gave me life?" (Rui a Papuhi).

"What is mine? / What is my identity?" (Patrick Amaru).

Poets, essayists, novelists, all have done their best to raise the consciousness of their compatriots and propose the possibility of a different future. But what comes now after literature, after the men's shame and the women's anger? Politicians, unavoidably. There was once a point of convergence, with Poovanaa Oopa, then at the centre of Tahitian politics. It coincided with a poetic outburst reminiscent of the old epic chants, reminded Tahitians of their identity, and celebrated the homeland and its culture.

In spite of the frequent political upsets between the two leading parties, there is little that is new today in Tahitian politics. What may each time be seen as a choice between the dizzying possibilities of radical change and the economic benefits derived from the status quo with France is, both directly and indirectly, at the base of modern Tahitian politics. In fact, it has been simmering for nearly three quarters of a century, sometimes reaching almost the boiling point, ever since Pouvana Oopa came back from the war and created the *Rassemblement des Populations Tahitiennes (RDPT)* out of the post-war frustrations. His party's political platform sought a more liberal constitution for Tahiti than existed at the time: putting native people to work in the public service, establishing agricultural and commercial cooperatives, creating adequate working conditions, and relaxing the exclusive economic ties with France. Today, these apparently modest and sensible desires no longer seem to be enough.

In the past, Tahitians were given the opportunity to decide on their fate through referenda, even if the options were outlined in a directive manner, such as the 1958 referendum on autonomy, where the choice and its consequences were clearly stated: Vote NO and you will belong to the big family of France[28], vote YES and you are on your own. Since the early success of Poovanaa Oopa, Tahiti has essentially been wavering between these two potential futures. Politics have really changed little and we see the same old opposition between the proponents of these two positions. Today, they are Oscar Temaru, the long-standing, pro-independent, and recently defeated leader, and the recently elected Gaston Flosse, who is said to want the tricolour flag of France to continue flying forever over the presidential palace in Papeete.

There is little that is new here, only the names occasionally changing. The autonomy movement never gets the majority that will convince even France that Tahiti deserves better, more, or perhaps something totally different. The vote on the next referendum could show whether Tahitians have come of age.

There are independent states in the South Seas, as the United States, Australia, and New Zealand have dismantled their colonial systems. France has not followed suit in French Polynesia, choosing rather to incorporate the islands into the French administration system. For Polynesians, there are enormous economic advantages attached to this situation, and a significant proportion of them are determined to keep the status quo.

The Economist of May 25, 2013 makes much of the "colonial yoke" to which French Polynesia is subjected in an article reporting on a United Nations' recent decision to reinstate (at the urging of Oscar Temaru, within hours of his losing the election the same month) the territory on its list of "non-self-governing bodies." Created in 1946, the list included territories that the colonial governments in charge had described as "dependencies." In the ensuing years, these dependencies sometimes changed their status, ranging from annexation to independence. Others remain on the list, often small islands with few resources of their own finding it to their advantage to continue with their connection to a larger nation.

Such is not the case in French Polynesia, and the territory is in a difficult position. So far, any time the pro-independence movement has made some progress in Tahiti, threatening to sway the voters, the less than subtle reaction from France has been to imply a withdrawal of support. However, Tahiti is very far from the Métropole, has not been pulling its economic weight, and is something of a financial burden to France.

THE NEW ARCADIA: TAHITI'S CURSED MYTH

In Tahiti, the eyes are at the moment on New Caledonia and its attempt at sorting out its own status. There, the support for independence is reported to be around forty percent, which would not normally be enough to succeed in a referendum. However, among the remaining sixty percent are the numerous French workers who hold temporary functions in New Caledonia. It has been recently agreed that the next referendum (to take place before 2020) would only be voted by those who have resided in the territory for twenty years, thus excluding all the short-term residents, originally from France. Whether their numbers are such as to weigh in the balance of votes is to be seen, but Mr. Temaru of Tahiti is closely watching the events in New Caledonia.

However, what should not be forgotten in this close watch by Tahiti is that New Caledonia is rich in mineral deposits and particularly in nickel, while Tahiti lacks any substantial resources, apart from tourism and copra.[29] In short, the population seems to be pulled between those who seek autonomous status for Tahiti, whatever the economic costs, and those who have opted, whatever the cultural costs, for the convenience and the benefits resulting from the ties with France, with presumably a good number hovering somewhere in the middle, trying to accommodate both aspirations. Or, in a few, odd cases, attempting to fly solo.[30]

The numbers have never been enough so far to win autonomy through a referendum. There is the same constant desire for a few politicians and their followers to divest themselves of the imposition of their post-colonial status and to achieve various forms of autonomy. In doing so, they have the backing of activist writers and thinkers, and most of the rural population. Others continue to be convinced that Tahiti's interests lie in keeping close ties with France, very likely people whose personal economic achievements have found a solid base in the status quo, or who genuinely believe that Tahiti would not be able to survive economically after over two centuries of dependency.

When it comes to another referendum, if such should be proposed, the example of New Caledonia in exempting temporary residents from the vote would be a good one to follow, to ensure that the permanent population is the only one to decide on its future. As things evolve, and with the option of autonomy still to be debated and voted upon, the attitude of

France towards the possibility of self-government in Tahiti and her islands, will have much import.

The French administration has created in Tahiti the infrastructure for the provision of modern services. A well educated, often *demi* working force would presumably have the skills to continue the administration of the territory and the management of its natural environment and resources, as well as ensuring the delivery of educational, medical, and social services. Other *demis* would continue to constitute the opposition, as they have so far done, not against France this time, but against themselves, surely the best sign of political maturity they could demonstrate.

On the other hand, administration by the privileged *demis* may not be welcomed by all. We have already read Spitz' dismissal of them in her novel's Epilogue. Mateata-Allain, as well, sees these "tutored half-caste Ma'ohi/Europeans" as perpetuating "colonial attitudes and policies." What she mostly resents is the contrast they present with the indigenous group, the majority, who is left "undereducated, under-employed, and marginalized from the high standard of living enjoyed by minority Europeans and the new race of French-fluent, half-caste Polynesians, the *demis*" (2003:3). Perhaps, like many Tahitians, she fears that the transition from the French to the *demis* would bring many problems, some perhaps worse than the existing ones, as they would be inflicted by their own half-caste people who feel themselves to be superior to the ordinary native population. The fact that many thinkers are themselves *demis*, but with different aspirations for their country, may not be reassuring enough for her and others like her.

It certainly would not be an easy transition, particularly for a nation devoid of natural resources and having to resort to tourism for much of its financing. We have seen that the success and profitability of tourism are mostly due to the effort of the administration, as Tahitians on their own would prefer to promote cultural activities and renewal for themselves. However, it may not take much for them to accept the necessity of taking tourism most seriously, as one of the islands' main source of revenue.

Not everyone would be willing to accept a change that would reduce their quality of life, if modern facilities are what they seek. For them, the return to a simpler life resulting from independence would not be seen as a step forward. If one believes previous referenda, this group is essentially urban and salaried, and would be deeply affected by any substantial changes in Tahiti's relations to France.

Yet, given conditions the world over and their impact on Tahiti (the general world economy—more particularly the current French economy and the cost of supporting services overseas—and the uncertainty of maintaining world tourism to its previous level) Tahitians might find themselves caught in the same precarious situation whether they stayed as part of France or not.

Today, the general thought worldwide is towards attempting to reduce and simplify, and towards achieving some form of self-sustainability. Those are economic considerations brought about by the economic downturn of the last few years. Spitz is open about it when she says, "We rely on the crash of the euro and the Occidental system to be forced to return to this [ancient] type of life" (Pernoud, 2012). While not yet universally accepted, these tendencies may no longer constitute an option. In Tahiti, those favourable to a more self-sustainable and modest way of life are the same people who, caught in a web of idealism, would see in political and economic autonomy the possibility of restoring the integrity of their Ma'ohi identity.

Unity could become problematic and existing divisions would very likely intensify if Tahiti and the other Polynesian islands were brought together into the same mold in spite of their different histories. Moreover, their urban and rural populations have different means, needs, aspirations, and religions, not to mention different political concerns.

For resentful Frenchmen, loath to abandon the islands and what they may have seen as a commitment to their *mission civilisatrice* among the islanders (as well as providing a pleasant if temporary appointment for civil servants and others), the question may simply be whether these islanders "can go back to the farm now that they've seen Paree"—Paree being all the benefits derived from Tahiti's current economic, social, and cultural relations with France. It is a crass observation, not very generous, and crudely formulated, yet one that holds much truth when we consider the upheaval that would constitute autonomy for Tahiti and her islands. But Tahitians are no strangers to upheaval; and, this time it would not be imposed upon them by others.

Louis-Antoine de Bougainville spent nine days in Otaheite, long enough for him to create and help propagate the myth of an ideal land, where nature and people conspired to reach a near-perfect balance, and where women were beautiful and generous with their lovemaking. He had ignored the constant intertribal strife and what would have seemed to him the savage practices that had prevailed in this paradise. Almost two and a half centuries later, his vision of paradise, which he shared with most of Europe, is still bringing to Tahiti people chasing the same dream. The question is whether contemporary Ma'ohi can adapt to the reality of an independent Tahiti in the twenty-first century, and accept to live within its constraints.

The Polynesians Bougainville met are no longer, and neither is the world they knew. They were forced within a few years into a maelstrom of change so drastic as to make one marvel at their survival into our century. These new and different people, even if they draw their strength from their mythical past are facing a future that could bring essential changes into their lives—anxiously desired by some, forcefully opposed by others.

The double myth is still alive: a self-sustaining life in harmony with nature (which modern Europeans probably doubt could be achieved again by the islanders after so many years of dependency) and the amorous *vahine* (which modern Polynesians resent because it debases them, yet exploit because it helps draw tourists). The latter would very likely be easier to maintain as a touristic and economic necessity, while the former would need the good will and patience all Polynesians would have to muster.

Renato Poggioli's reflection on the pastoral impulse may apply as strongly to some modern Tahitians as it did to eighteenth-century Europeans or to anyone with nostalgic desires for an imagined past. The "pastoral embrace, that is the hankering for an easier form of the world, is never destroyed" (Fussell, 1988:151). This universal need, both "longing and wish fulfillment," is changing its external shape over time. It is natural enough that, for many Tahitians, it should be rooted in the Otaheite of old. Unfortunately, Poggioli also reminds us that this universal search, so deeply entrenched in the human mind, is little more than a fallacy.

However, in some legends, the heroes, having suffered much and gone through dreadful tests and ordeals on their quest, are finally allowed to come home, far better men. These tales are about passage and growth, and are charted through the three main events of departure, initiation, and return. Should the Ma'ohi be able to make this awesome journey from contemporary Tahiti back to their original source, they should also remember that the lesson of the legend is to retain in their new lives the wisdom learned during the hardships of their quest.

NOTES

1 From a guidebook I found in Havana in 1992 that also reported a statement attributed to Hatuey, a local Indian chief who had previously fought the Spaniards on Hispaniola: "These Europeans worship a very covetous sort of God. They will exact immense treasures of us and will use their utmost endeavours to reduce us to a miserable state of slavery or else put us to death."

2 Many of the historical sequences in this book are based on Robert W. Kirk. *History of the South Pacific since 1513* and some of his own sources: Gavan Daws, *Shoal of Time. A History of the Hawaiian Islands*; Ernest S. Dodge, *Islands and Empires: Western Impact on the Pacific and East Asia.*; J. C. Furnas, *Anatomy of Paradise*; Allan F. Hanson. *Rapan Lifeways: Society and History on a Polynesia Island*; David Howarth, *Tahiti: A Paradise Lost*; Richard Landsdown, *Strangers in the South Seas: The Idea of the Pacific in Western Thought*; Frank Sherry, *Pacific Passions: The European Struggles for Power in the Great Ocean in the Age of Exploration.*

3 Most of the mutineers were recaptured and kept on the *Pandora* in a fetid and harsh prison. The ship was wrecked on her way back, with loss of life for both crew and mutineers; however, her three boats eventually succeeded in reaching Timor, with the promise of English justice for the surviving mutineers at the end of the voyage home.

4 After Mexico established footholds in the Philippines, Peru took over from Spain the continuing exploration of the Pacific. See Leigh

Reyment, Alvaro de Mendaña de Nehra. http//www.nl/tue/~engels/discovery/mendanha.heml

5   There have been several experiments to replicate the migrations of ancient Polynesians and shed another possible light on their origins—not from the east, as commonly thought, but rather from the American coast. Some tested these theories, such as Thor Heyerdahl sailing on the *Kon-Tiki* raft in 1947. This serious attempt met with some reservations—or even outright mockery. (Paul Theroux, in *The Happy Isles of Oceania: Paddling the Pacific*: "In a lifetime of nutty theorizing, Heyerdahl's single success was his proof in Kon-Tiki, that six middle-class Scandinavians could successfully crash-land their raft in the middle of nowhere"). Danielsson, the anthropologist and traveller I often quote, was one of these six Scandinavians. Eric de Bisshop, another sailor on board the large raft *Tahiti Nui*, successfully showed in 1956 that Polynesians could have reached the coast of South America in ancient times and been able to come back. Unfortunately, he died at sea on his return journey to Tahiti

6   The Hawaiian Encyclopedia. (*Rediscovering the Past: The Revival of Polynesian Voyaging Traditions*) provides a description of how ancient Hawaiians trained and developed their remarkable powers of observation On Mount Moaulaiki, ocean navigators were trained in the art of celestial navigation from visual clues: a view of other islands and the observable currents in the channels between them—of particular interest since currents visible on the ocean's surface are created by winds, water pressure, and temperature.

7   John Marra, a gunner's mate on the *Resolution*, a man of poor repute but with a lively pen, was among the first to publish a journal of the voyage to Tahiti. Not sanctioned by Cook, sometimes referred to as "illicit," it was received with great interest, first published in London in 1775, then in Dublin in 1776, with German and French translations published in 1776 and 1777. See Tom Ryan's "Blue-Lip'd

Cannibal Ladies: The Allure of the Exotic in the Illicit Resolution Journal of Gunner John Marra." http://researchcommons.waikato.ac.nz/bitsream/handle/10289/3203/Blue-Lip'd-pdf?sequence=1

8   Francophone singers seem to have had a particular affinity with the islands. Joe Dassin, first intending to spend only three days in Tahiti, stayed three weeks before his contracts forced him to return home, then eventually came back to die there. Antoine titled one of his albums "Tahiti and her Islands: Return to Paradise," and Jacques Brel sang in "Marquesas" *Les femmes sont lascives au soleil redouté."*

9   Although she may have come close to it. In his *Supplément au Voyage de Bougainville*, Diderot describes the scene. After being made welcome by the Tahitians, the Europeans suddenly heard the cries for help from one of the officers' servants. "Young Tahitians had thrown themselves on him, stretched him on the ground, undressed him, and were disposed to pay him their compliments... This servant was a woman disguised as a man. Ignored by the crew during the whole of the crossing, the Tahitians guessed her sex at first sight." She was rescued. However, Dunsmore (2002:97) believes that the natives may have only been driven by curiosity and the desire to see the white bodies hidden under European clothing, rather than by the intent to rape poor Jeanne, since the male cook had previously suffered the same fate without any consequences. Apparently, the first Tahitian on board, Ahutoru, had immediately recognized her as a woman in spite of her disguise and had thought her a *mapou* (transvestite). The "attack" may simply have been a desire to check whether this was true. The French assumed on the basis of this recognition that the Tahitians had an uncanny ability to discern—by smell, as was suggested, or by some other means—the sex of women, thus attributing to them extraordinary abilities in such matters.

10 I am reminded of an unspoken, and certainly less formal, tradition in the French bourgeoisie of earlier times which held that fathers

would have their sons initiated by a friendly prostitute at the brothel they also patronized. There was also the tacit and never acknowledged arrangement some of the boys' mothers might have had with one of their younger women friends tending to the discreet *dépucelage* of their teenaged sons.

11   In this section, I am indebted to Robert W. Kirk's *History of the South Pacific since 1513,* and to his own sources: Gavan Daws, *A Dream of Islands* and *Shoal of Times. A History of the Hawaiian Islands;* Ernest S. Dodge, *Islands and Empires;* John Garret, *To Live Among the Stars: Christian Origins in Oceania;* C. Hartley Grattan, *The South Pacific to 1900;* K. R. Howe, *When the Waves Fall: A New South Sea Islands History from First Settlers to Colonial Rule;* Maretu, *Cannibals and Converts: Radical Changes in the Cook Islands,* and to the writers of contemporary religious tracts in 1888.

12   In *Breadfruit,* Vaite relates two myths, both based on love, about the origin of the all-important breadfruit and why coconut have three dots. In the first instance, a man transforms himself into a breadfruit tree so that his woman and children have something to eat. In *Frangipani,* she relates the legend of Princess Hina, ordered by her father to marry a prince she finds ugly (he is in fact an eel). She decides to have him killed and appeals to the God Maui, who captures the eel and cuts him into three slices. The Princess is then told to bury the eel's head, wrapped in tapa cloth, in the familial *marae*. She forgets, and a number of events ensue (including the sprouting of a strange looking tree resembling an eel), during which she hears a voice crying out "Princess Hina, you are going to look into my eyes, you're going to kiss my mouth. You're going to love me." Finally, in the middle of a terrible drought afflicting the island, she returns to that strange looking tree, and, peeling its fruit, she sees three dots appear and remembers the eel's words. She presses her mouth to the eel's mouth and realizes how much he had loved her. Other myths explain the islands' geography. For instance, Tahiti-Nui and Tahiti-Iti were formed when a great fish swam from the sacred waters of Havai'I, to create the lagoon now

shared by Raiatea and Taha'a. Another version explains that the two islands were originally one, until a lovelorn giant, rejected by his lover, angrily smashed his fist into the island, splitting it into two parts.

13 Loti supports Gauguin's contention and relates the arrangements made with the family of Rarahu, his "little girl." "Rarahu's old foster-parents, whom I at first feared to grieve, had ideas on such subjects.... They said that a great girl of fourteen was no longer a child, and was not created to live alone. She did not walk the streets of Papeete and that was all they asked of her propriety" (Loti, 1986:27).

14 Were those ships, Boenechea's *Aguila* and *Jupiter*, sailing in the islands between 1772 and 1775? The religious faith of the Spanish captain and his sponsor, the viceroy of Peru, might not have been enough to guarantee that venereal diseases were not present on board, but Boenachea had expressly forbidden all the crew, himself included, to have sexual relations with Tahitian women. This abstention certainly surprised the Tahitians, accustomed to the behaviour of English and French crews. Moreover, the natives had observed with much interest the use of a casket in Boenechea's funeral in Tahiti in 1775. Thus, it is surprising that they only based their recognition of a Spanish identity on seeing a flag and bits of clothing shown by Cook.

Yet, the natives blamed their Spanish crews. There had been other ships. The *San Lorezo* and the *Santa Rosalía*, under Captain González de Haedo, landed on Rapa Nui in November 1770. Apart from these and Bonachea's four vessels, there does not seem to be any other mention of the Spanish presence in Tahiti or surrounding islands in the eighteenth century. But, perhaps those were not the first to have reached Polynesia. Danielsson, who treats the visit of Mendaña to the Marquesas in the 1595 with great contempt, writes about the skirmishes between the natives and the Spaniards, whom he describes as brutal and greedy. Mendaña then moved on to Vaitahu, where his men found local women very attractive, with "pretty legs and hands, fine eyes, pleasing faces, slim waists, and many of them were prettier than the women in Lima who are renowned for their beauty," wrote a Spanish chronicler. Mendaña, finding neither gold nor precious stones on the Marquesas, sought his fortune further westward. Unfortunately, according to Danielsson, "at Vaitahu, which he had rechristened Valle de Madre de Dios, he left syphilis and 200 dead as a reminder of his visit" (1957:36). The 200 dead likely occurred during the fighting Danielsson describes earlier, but he does not give any source to support his allegation that Mendaña's men introduced venereal disease to the Marquesas, from where it could have spread to other islands over time. Although the Spaniards may have behaved with more circumspection than the English and the French, there were rumors in Tahiti of a certain Martimo (Salmond suggests "Maximo"), a member of the crew from Lima, who had learned the natives' language, participated in their rituals, and had had hearty commerce with their women. He seemed to have been part of a group consisting of two priests and their servant, who built the Catholic mission that was to be abandoned a few months later in 1775, after Boenachea's death.

15   Adding considerably to the sorrow and the sense of injustice those events caused, carelessness was sometimes also suspected, such as in British Columbia, during the dreadful smallpox epidemics that killed perhaps a hundred thousand natives during 1862-63. Blankets contaminated with smallpox had been distributed at a gathering place and followed their new owners back to their own villages. "The smallpox in its most virulent form has attacked all the tribes, more or less, on the coast," wrote a Victoria newspaper. While some vaccine, administered by two Anglican ministers, had been made available in Nanaimo, the rest of the people faced a terrible and lonely fate. The local press *(The Colonist* and the *Victoria Daily)* followed with fascinated horror the progress of the disease that would have been familiar to Polynesians (Swanky, 2012:294, 297).

16   Many examples exist of this pattern. For instance, with the Montagnais in Canada, these changes came early on. A populous nation in Quebec, settled in the Saguenay region, the Montagnais experienced

the typical process to which the Tahitians were later exposed. The first encounters had been with French cod fishermen and whalers in the fifteenth century, who were followed in the next century by French traders. These were soon providing lard, tea, cloth, and weapons in exchange for the Indian trappers' furs. The Jesuits established their first mission there in 1632, but there had already been a significant change in local customs by then. "By 1623 the Montagnais were no longer manufacturing birchbark baskets, clay pots, and stone adzes. Instead, they were purchasing copper kettles and iron axes... They also preferred French clothing, preferring the wool garments more practical than skin ones." We should remember that had been recently still using pelts and bones to make clothes and weapons. The trading had also extended to other merchandise, as legitimate traders were not the only ones to arrive on the scene, and as early as 1620 illegal traders had been providing the Montagnais with guns and ammunition. Weapons were not the only merchandise on offer. "Wine and liquors were supplied to the Indians by the independent traders who operated along the lower St Lawrence... Among the Montagnais, alcoholic beverages were probably initially valued for their hallucinogenic properties and viewed as a way of communicating with the supernatural. Other Canadian tribal nations had the same needs and traded with the French. "Huron traders were primarily interested in obtaining metal cutting-tools. In particular, they wanted iron knives of all sizes, awls, axes, and iron arrowheads... They also sought copper and brass kettles." Similarly, modern archeologists have found a marked decline in stone tools in the Iroquois sites they examined. Interestingly, little changed was observed with women's tools, "either because of the greater conservatism of women or because male traders chose items that were useful to themselves rather than to women" (Trigger, 1985: 203-205, 209-10).

17 This section is based for the most part on K.R. Howe, *Where the Waves Fall: A New South Sea Islands History from First Settlers to Colonial Rule*: Robert Kirk, *History of the South Pacific since 1513;* Colin Newbury, *Tahiti Nui: Change and Survival in French Polynesia, 1767-1945;*

David Stanley, *Tahiti;* and William Tagupa, *Politics in French Polynesia, 1945-1975.*

18 Most French colonies faced the same dilemma: follow General de Gaulle or Marshal Pétain. Most rallied to de Gaulle. A certain General Leclerc (an alias to protect his family), after escaping from a German prisoner camp, had fled to London, from where he had been sent by de Gaulle to constitute a French army in Africa. He soon formed a ragtag group of some three thousand, made up among others of *tirailleurs sénégalais* (Senegalese skirmishers), a small armoured unit, and some pilots from Brittany who had rallied with their elderly planes. This army started from French Equatorial Africa, marched from Chad to Tripoli, and met there with General Montgomery's Eighth Army. Leclerc was then ordered back to London to organize the French 2nd Armoured Division *(2ème Division Blindée)* and prepare for the Normandy landing.

19 Those three hundred volunteers from Tahiti were joined in Noumea by three hundred New Caledonians. From Noumea, they sailed to Australia and then to Suez. They joined with two French Foreign Legion battalions, another colonial battalion from Chad, and a marine infantry contingent to constitute the 1st Free French Division under General Koening. They would be among the 3,700 French troops who held out in May–June 1942 against 35,000 Germans and Italians commanded by Rommel. Against two Panzer divisions, the Germans' 88 mm guns, and the support of the Luftwafe, they were besieged for two weeks in the Lybian desert at Bir Hakeim, until they could successfully rejoin the British troops. Nominally a part of the 7th British Tank Division (the "Desert Rats"), they were known as the Free French Brigade. Among those recognized for extraordinary bravery was a Tahitian, Jean Tranape (Miller, 2012).

20 Sources for Nuclear testing in Polynesia are:

- CNN. "Fifth French nuclear test sparks international outrage." December 28, 1995. http://www.cnn/World/9512/france nuclear/;
- Planet Canterbury from International Physicians for the Prevention of Nuclear War (IPPNW) Australia. "Health in French Polynesia – The Effects of French Nuclear Testing. http://www.Cyberplace.org.nz/peace/nuchealth.html;
- "Fifth French nuclear test sparks international outrage. *World News*. Dec. 28, 1995;
- Lucas, Clarize. "The Nuclear Colonialism" (Translated from the French). *Poison Fire, Sacred Earth: Testimonies, Lectures, Conclusions*. The World Uranium Hearing, Salzburg. September 17, 1992. pp. 212-213.http://www.ratical.com/radiation/WorldUraniumHearing/index.html
- Paul Benkimoun, "Les essais nucléaires polynésiens responsables de cancers thyroïdiens." *Le Monde*, Aug. 3, 2006.

21 Very few of us have forgotten the sinking in Auckland, New Zealand, on July 10, 1985, of the *Rainbow Warrior*, the Greenpeace ship ready to lead the opposition to the French tests. The vessel was supposed to be empty, but a photographer on board was killed in the explosion. A large fine was paid and the French Defense minister resigned. The two agents responsible for the sinking were sentenced to ten years in jail, of which they only served two.

22 French history books used to start with *"Nos ancêtres, les Gaulois."* Our ancestors, the Gauls, who were described as tall, blond, with blue eyes, and fearing nothing more than the sky falling over their heads. The same books were used throughout the former French Colonial Empire. From Senegal, to Indochina, to Madagascar, to French Polynesia, we (I, in Morocco) learned from the same curriculum and were exposed to the same fiction. It may well have blended with the stories the child from Tonkin, Côte d'Ivoire, or Tahiti was told at home, based on his (for in those days the colonial school child would most likely have been a boy) culture's oral history. However,

once grown up, he may well have started resenting what he had been taught as a child against all evidence. This common education was part of the intended civilizing effect of colonialism, and France was only one of those European countries that spread their domination and administrative savoir-faire well across their borders and their own continent.

23 In Commerson's appreciation of the Tahitian language (Montessus, 1889:58-59), we see a reflection of the values attached to being a Noble Savage, in contrast to the overly refined European languages. Commerson praised this "very sonorous, very harmonious language." With only about four or five thousand words, and no syntax whatsoever, "it is enough to enable them to express all their ideas and all their needs. [It has] a noble simplicity which, by not excluding tonal modifications, nor the pantomime of passions, saves them from this vain bathos we call the richness of language through which we lose in the labyrinth of words the precision of our perceptions and the immediacy of our judgement."

24 For the section on Tahitian literature, I have mostly relied on the following sources: Sylvie André. "La Littérature polynésienne en français"; Flora Devantine. "Oralité-Oraliture-Littérama'ohi"; Daniel Margueron. "Tahiti ou l'atelier d'une invention littéraire. Présentation des littératures en Polynésie française"; Kareva Mateata-Allain. "Orality and Ma'ohi Culture" and "Ma'ohi Women Writers of Colonial Polynesia: Passive Resistance toward a *Post*(-)colonial Literature"; *Polynesian Songs and Chants (Society Islands);* Bruno Saura. "Quand la voix devient la lettre: Les anciens manuscripts autochtones *(puta tupuna)* de la Polynésie française"; Cadousteau Vaihere. "Le 'Oreo: Le renouveau d'un art antique oratoire."

25 The custom persists. In 2014, I had a long conversation in Papeete with a man who had been intrigued to find out I lived in British Columbia, having recently seen a documentary on the salmon run

in our BC rivers. After I had described to him the fish cycle of life and death as well as the fishing techniques of the eagles and the bears—almost mythical animals to him—we turned to our respective families. A middle aged man, a *demi* many times over via Normandy and Noumea, he explained that when his last daughter was born, his younger brother had exclaimed that he "had to have" her. The young man adopted her and, as my new friend explained, since they all lived together, the little girl had her two *papas* under the same roof. He, too, had happily adopted a cousin's child.

26 It was probably the seed of the *noni* tree, one of those "miracle" fruit reputed to be a panacea of sorts and to have considerable healing virtues—much as argan oil in Morocco: it too has undergone the same transformation. At first, merely a seed from which oil was extracted and locally appreciated in cooking or used to soften the skin, it has been taken over by modern marketing. Today, the list of its qualities seems nearly endless, and it is sold at great cost everywhere in the world (sometimes cut with the cheaper olive oil), to be taken internally for health or used externally for beauty. Both the argan and the *noni* are good sources of revenue for their respective countries.

27 We can see the old method still being applied in Hawaii, in a 2014 documentary, "Ancestral Ink" *(Pacific Heartbeat),* where the Hawaiian tattoo master, Keone Nunes, practices the traditional skill he learned from an older Samoan master. He uses the tools he made or inherited from his master: the *moli* (a sort of fine-toothed "comb" that applies the ink into the skin, following the design drawn on the skin) and the *hahau*, the hitting stick. He does not touch the person he tattoos, as the skin is stretched by one or two other men. As he explains, "The tools will let me know whether or not the skin is doing well." It is a very noisy affair, with the stick beating fast and strong on the *moli*. No doubt, bearing the long painful process with patience and in silence is a matter of pride.

28 The French Community (replacing the French Union) was no longer in 1958 what was once the vast French Colonial Empire. France was still reeling from her defeat in the murderous War of Indochina, and was engaged in an equally murderous war with Algeria that would only end in 1962. Furthermore, there were already intimations of unrest in her African colonies, which would all be gaining their independence through a number of referenda in the 1960s.

29 New Caledonia is in the enviable position of having the world's fourth largest deposit of nickel in the north of the country and the nearby Loyalty Islands. Following the threat of civil unrest, the Kanak *indépendantistes* were given control over the exploitation of the Koniambo Mountain by the French government in June 1988. Today, thanks to two individuals, André Dang, a long-time New Caledonian anticolonialist and president of the *Société Minière du Sud Pacifique*, and Paul Neaoutyine, an *indépendantiste* and president of the North Province, the Kanaks are able to negotiate on their own terms (their ownership always remaining at 51%) with several multinationals for the exploitation of their nickel. They bring in the resources, and the foreign companies (e.g. Falconbridge, Xstrata Nickel, Posco) provide funding and technology. They agree to their minority share because of the richness of the deposit and the unusual forty-kilometre proximity between the mine, the factory, and the deep-water harbour. With this industrial exploitation and six thousand workers from twenty-eight different nationalities, the Kanaks (previously fishermen and farmers) are wary of the threat to their environment and their culture. Nevertheless, it is a risk they are prepared to take to achieve economic and political independence.

30 Occasionally, those who find the status quo galling, attempt to create a new party with a small band of followers. Thus, the new republic of [Repupirita] Hau Pakumotu, was proclaimed in November 2009 in Moorea and sanctified with a ceremony on the Halora *marea* in New Zealand. Its president, Athanese Terii, soon decided to become king. Without a land to rule, he has nevertheless appointed a full

government. Throughout 2010, the new president-king was much in the news and the Papeete press kept everyone posted on the antics of the one they called the "Puppet King" or the "Mad King." Hau Pakumotu may not have made too many inroads into the local political structure, and it would be only too easy to mock President Terii/King Pakumoto's attempts at creating a new state, but the genuine desire shown by some of his followers for something new, something genuinely ma'ohi, is not to be ignored.

GLOSSARY

Aparima: danse telling a story with the hands

Ari'i: high chief, head of clan or tribe

Arioi: religious society of entertainers, travelling among the Society Islands

Breadfruit: large round fruit with starchy flesh grown on the *uru* tree

Copra: dried coconut meat

Demi: Tahitian issued of a Tahitian mother and European/American father

Fafa: cooked taro leaves

Fiu: fed up, bored

Hiro: god of thieves

Kaina: local person

Lagoon: expanse of water surrounded by a reef

Lapita: pottery of the ancient Polynesians (1600-500 B.C.)

Le truck: truck with benches, commonly used as transportation for people

Mahu: transvestite

Ma'ohi: indigenous Polynesian

Mama, Meme: familiar address for mothers and grandmothers

Marae: temple consisting mainly of an open space and a stone platform

Monoï: perfumed coconut oil

Motu: low-lying island

Oro: god of war

Pandanus: slender tree (pine) whose leaves are plaited into hats or mats

Papa'a or popa'a: European

Pareu: wrap around skirt (women) or loin cloth (men)

Peretane: British

Pito: navel, umbilical cord

Reef: coral ridge near the ocean surface

Tahua: skilled artisan, healer, or priest

Tamure: very fast erotic dance

Tane: man

Tapa: cloth made from the bark of the mulberry tree

Tapu: taboo, set apart, forbidden

Taro: root vegetable, staple food of Polynesia

Tavana: chief

Tiare: national flower of Tahiti

Tiki: human-like figure representing an ancestor, used as protection

Tinito: Chinese

Tumu: tree

Unu: carved panel displayed on the marae

'Uru: breadfruit tree

Vahine: woman

BIBLIOGRAPHY

Katokamanava. Myth, History and Society in *the Southern Cook Islands*. 1991.

Alloula, Malek. *The Colonial Harem* Minneapolis: University of Minnesota Press. 1986.

Amaru. G. "Tahití hubiera podido ser española." *Revista Vivat Academia*. No. 71. Diciembre 2005-enero 2006.

André Sylvie. "La littérature polynésienne en français." In Jean Bessière and J. M. Moura, eds. *Littérature postcoloniale et francophonie*. Paris: Champion. 2001. http://www.lehman.cuny.edu/ile.en.ile/pacifique_litterature.html

Anon. "The French in the Pacific" (pp. 8-13) and "Jottings from the South Pacific." (pp. 396-97). *The Sunday at Home. Family Magazine for Sabbath Reading* London: Religious Tract Society. 1888.

Banks, Sir Joseph. *The Endeavour Journal of Sir Joseph Banks*, by Joseph Banks, from 25 August 1768-12 July 1771. Sydney: The Trustees of the Public Library of New South Wales. 1962. http://gutenberg.net.au/ebooks05/0501141h.html.

Barrow, Terence. *The Art of Tahiti and the Neighbouring Society, Austral and Cook Islands*. London: Blacker Calman Cooper Ltd. 1979.

Beaglehole, J.C. (ed.). *The Journals of Captain James Cook*. Cambridge: Cambridge University Press, for the Hakluyt Society. 1955.

-------- (ed.) *The Journals of Captain Cook: The Voyage of the* Resolution *and* Discovery *1772-1775*. Cambridge: Cambridge University Press, for the Hakluyt Society. 1969.

-------. *The Life of Captain Cook*. London: Adam & Charles Black. 1974.

Bennett, Frederick Dobell. *Narrative of a Whaling Voyage Round the Globe, from the Year 1833-1836, Comprising Sketches of Polynesia California, the Indian Archipelago, Etc.* London: Richard Bentley. 1840.

Bernardin de Saint-Pierre, Jacques-Henri. *Journey to Mauritius* (1773). Northampton, MA: Interlink Books, 2003.

Bligh, William. *A Voyage to the South Sea*. London: George Nicol. 1790.

Bolyanatz, Alexander H. *Pacific Romanticism. Tahiti and the European Imagination*. Wesport., CT: Praeger Publishers. 2004.

Boswell, James. *The Life of Samuel Johnson*. New York: Alfred A. Knopf. 1962.

de Bougainville, Antoine. A *Voyage Round the World Performed by Order of His Most Christian Majesty, in the Years 1766, 1767, 1768, and 1769*. London. 1772.

Brami-Celentano, Alexandrine. "Le renouveau identitaire de la jeuness à Tahiti. La culture et l'identité *ma'ohi* en question." From "La jeunesse à Tahiti: renouveau identitaire et réveil culturel," *Ethnologie française*, 2000-4.

Braudel, Fernand. *The Structures of Everyday Life*. London: Collins 1982.

Brendon, Piers. *The Decline and Fall of the British Empire, 1871-1997*. London: Vintage Books, 2007.

Brettell, Richard, Françoise Cachin, Claire Frèches-Thory, Charles F. Stuckey. *The Art of Paul Gauguin*. National Gallery of Art: New York Graphic Society Books. 1988.

Burling, Robbins. *Man's Many Voices. Language in its Cultural Context*. New York: Holt, Rinehart and Winston, Inc. 1970.

Burney, James. *With Captain James Cook in the Antarctic and Pacific. The Private Journal of James Burney, Second Lieutenant of the "Adventure" on Cook's Second Voyage, 1772-1773*. London. 1775.

Cameron, Roderick W. *The Golden Haze. With Captain Cook in the South Pacific.* London: Weidenfeld and Nicolson. 1964.

Chaze, Michou. *Vai la Rivière au Ciel sans Nuages.* Papeete: Cobalt/Tupuna/Les Editions de l'Après-midi. 1990.

Childs, Elizabeth C. *Vanishing Paradise. Art and Exoticism in Colonial Tahiti.* Berkeley: University of California Press. 2013.

Codere, Ellen, ed. *Kwakiutl Ethnography.* Chicago: Chicago University Press. 1966.

Cook, James. *Captain Cook's Voyages round the World.* London: W. Wright. 1834.

------ Captain Cook's *Journal During his First Voyage Round the World Made in H.M. Bark "Endeavour", 1768-71.* A Liberal Transcription of the Original MSS. With Notes and Introduction edited by Captain W.J.L. Wharton, R.N., F.R.S. London: Elliot Stocke. 1893. http://www.gutenberg.org/files/8106-h/8106-h-htm#ch2.

CNN World News. "Fifth French nuclear test sparks international outrage." December 28, 1995. http://www.cnn/WORLD/9512/france nuclear/

Corney, B. G., ed. *The Quest and Occupation of Tahiti by Emissaries of Spain during the Years 1772-6.* (3 vols.) London: Hakluyt Society. 1913-1919.

Danielsson, Bengt. *The Happy Island.* London: George Allen and Unwin. 1954.

-------*Work and Life on Raroia. An Acculturation Study From the Tuamotu Group, French Oceania.* London: George Allen and Unwin. 1956a.

------ *Love in the South Seas.* New York: Reynal & Company. 1956b.

------ *Forgotten Islands of the South Seas.* London: George Allen and Unwin Ltd. 1957.

-------- *Gauguin in the South Seas.* London: George Allen and Unwin Ltd. 1965.

Danielsson, Bengt and Marie-Thérèse. *Moruroa, Mon Amour*. Paris: Editions Stock. 1974.

Darwin, Charles. *A Naturalist's Voyage Round the World: The Voyage of the Beagle*. (First edition, May 1860). Chapter XVIII. www.gutenberg.org/catalog/world/readfile?fk_files=2966362

Daws, Gavan. *A Shoal of Time. A History of the Hawaiian Islands*. Honolulu: University of Hawai'i Press. 1968.

------ *A Dream of Islands: Voyages of Self-discovery in the South Seas*. Honolulu: Mutual. 1980.

A. Grove. *Hawaii and its People*. Honolulu: University of Hawai'i Press. 1968.

Dening, Gregory. "Ethnohistory in Polynesia. The Value of Ethnohistorical Evidence." *The Journal of Pacific History*. Vol. 1, 1966.

Denoon, Donald, ed. *The Cambridge History of the Pacific Islanders*. Cambridge: Cambridge University Press. 1997.

Devantine, Flora. *Tergiversations et rêveries de l'écriture orale: Te Pahu a Hono'ura*. Papeete: Au Vent des Iles. 1998.

------- "Oralité-Oraliture-Littérama'ohi." *Littérama'ohi: ramées de littérature Polynésienne, te hotu Ma'ohi*. No 2. 2009.

Diamond, Jared. *The World Until Yesterday. What Can We Learn From Traditional Societies?* New York: Viking. 2012.

Diderot, Denis. *Supplément au Voyage de Bougainville*. (First published in 1796). Paris: Gallimard. 2002. Also: In Libro Veritas (Collection Philosophie): http://www.bacdefrancais.net/diderot-supplement-voyage-bougainville-pdf

Dodd, Edward. *The Rape of Tahiti*. Volume IV of *The Rings of Fire*. New York: Dodd, Mead & Company. 1983.

Dodge, Ernest S. *Islands and Empires: Western Impact on the Pacific and East Asia*. Minneapolis. 1976.

Dodsley, Robert. "Characters. Genuine Account of Omiah, a Native of Otaheite, a new discovered Island in the South-Seas, lately brought over to England by capt. Fourneaux." *Annual Register, 1774.* London: 1774.

Doon, Donald, ed. *The Cambridge History of the Pacific Islanders.* University of Cambridge Press. 1997.

Driessen, Hank. "Outriggerless Canoes and Glorious Beings." *The Journal of Pacific History.* XVII. 1982.

Druett, Joan. *Rough Medicine. Surgeons at Sea in the Age of Sail.* New York: Routledge. 2001.

Dunsmore, John. *Monsieur Baret. First Woman Around the World (1766-68).* Auckland, NZ: Heritage Press Limited. 2002.

Elbert, Samuel H. "Chants and Love Songs of the Marquesas Islands, French Oceania." *Journal of the Polynesian Society.* Vol. 50, No. 198. Pp. 53-91

Ellis, Juniper. *Tattooing the World. Pacific Designs in Print and Skin.* New York: Columbia University Press. 2008.

Ellis, William. *An Authentic Narrative of a Voyage Performed by Captain Cook and Captain Clerke, 1776-1780.* London: G. Robinson. 1783.

------- *Polynesian Researches during a Residence of Nearly Six Years in the Society and Sandwich Islands.* Vols. I-IV. London: Henry G. Bohn. 1859.

Fidlon, Paul G. and R. J. Ryan, eds. *The Journal of Arthur Bowes-Smyth: Surgeon, Lady Penrhyn, 1787-1789.* Sydney: Australian Document Library. 1979.

Forster, George. *A Voyage Round the World in 'Resolution'.* Vols. I-II. London, 1777.

Fresche, Charles-Félix-Pierre. *Tahiti au nom du Roi: 1768, Bougainville débarque à Tahiti.* Paris: Nicolas Chaudun, 2007)

Furnas, J. C. *Anatomy of Paradise.* New York. 1948.

Fussel, Paul. "The Persistence of the Pastoral" and "Travel, Tourism, and International Understanding." In *Thank God for the Atom Bom and Other Essays*. New York: Ballantine Books. 1988.

Garrett, John. *To Live Among the Stars. Christian Origins in Oceania*. Geneva: World Council of Churches, in association with the Institute of Pacific Studies. University of the South Pacific. 1982.

Gauguin, Paul. *Noa Noa. The Tahitian Journal of Paul Gaugin*. San Francisco: Chronicle Books. 1994.

Gesner, Peter, ed. *A Voyage Round the World in His Majesty's Frigate* Pandora. Sydney: Hordern House, for the Australian Maritime Museum. 1998.

Gilbert, John. "Charting the Vast Pacific." *Pacific Voyages. The Encyclopedia of Discovery and Exploration*. Part II. Garden City, NY: Doubleday. 1973.

Grattan, C. Hartley. *The South Pacific to 1900*. Ann Arbor, MI. 1963.

Hale, John. *Age of Exploration*. New York: Time Incorporated. 1967.

Hambly, Wilfrid Dyson. *The History of Tattooing*. Mineola, N.Y.: Dover Publications, Inc. 2009.

Hanbury-Tenison, Robin. *The Oxford Book of Exploration*. Oxford: Oxford University Press. 2005.

Hanson, Allan F. *Rapan Lifeways: Society and History on a Polynesian Island*. Boston. 1970.

Hawksworth, John. *An Account of the Voyages undertaken by Order of His Present Majesty for making Discoveries in the Southern Hemisphere*. London. 1773.

Henry, Teuira. *Ancient Tahiti*. Honolulu, Hawaii: Bernice P. Bishop Museum. 1928.

Holland, Julian. "Lands of the South Cross." *Pacific Voyages. The Encyclopedia of Discovery and Exploration*. Part III. Garden City, NY: Doubleday and Company, Inc. 1973.

Holmes, Christian, ed. *Captain Cook's Second Voyage: The Journal of Lieutenants Elliott and Pickersgill*. London: Caliban Books. Revised edition, 1988.

Howarth, David. *Tahiti: A Paradise Lost*. New York. 1983.

Howe, Kenneth R. *When the Waves Fall: A New South Sea Islands History from First Settlers to Colonial Rule*. Honolulu: University of Hawai'i Press. 1988.

Igler, David. *The Great Ocean. Pacific Worlds from Captain Cook to the Golden Rush*. Oxford: Oxford University Press. 2013.

International Physicians for the Prevention of Nuclear War Australia. "Health in French Polynesia. The Effects of French Nuclear Testing." July 1995. http://www.cyberspace.org.nz/peace/nuchealth.html

Jacobs, Joseph. *The Story of Geographical Discovery. How the World Became Known*. University of Pennsylvania Electronic Classic Series Publication. 2004.

Kahn, Miriam. *Tahiti Beyond the Postcard: Power, Place, and Everyday Life*. Seattle: University of Washington Press. 2011.

Kamakau, Samuel M. and Z. Kepelino. "Hawai'ilo and the Discovery of Hawai'i." From *Kepelino's Traditions of Hawaii*. Honolulu: Bishop Museum. 1932.

Kawaharada, Dennis. *The Settlement of Polynesia*, Part I. http://www.paulwaters.com/migrate.htm

Kirk, W. Robert. *History of the South Pacific since 1513. Chronology of Australia, New Zealand, New Guinea, Polynesia, Melanesia and Robinson Crusoe Island*. Denver, Colorado: Outskirts Press. 2011.

von Kotzebüe, Otto. *A New Voyage Round the World in the Year 1823*. London: Colburn & Bentley. 1830.

Kroepelien, Bjarne. *Tuimata*. (First published in 1944). Including "Annexes" by Jean-Claude Teriierooiterai, Rolf du Rietz, and Daniel Margueron. Papeete, Tahiti: Haere Po. 2009.

Landsdown, Richard, ed. "Introduction" (pp. 1-27) and "Noble Savage" (pp. 64-72), in *Strangers in the South Seas: Idea of the Pacific in Western Thought*. Honolulu: University of Hawai'i Press. 2006.

Landsdowne, Marques de, ed., *The Queenery Letters, Being Letters addressed to her Majesty Mary Thrale*. New York: Farrar and Rinehart. 1934.

La Pérouse, Jean-François. *Voyage de La Pérouse Autour du Monde*. Paris. 1797.

Lavondès, Anne, "Les dessins de Philibert de Commerson et la Culture Tahitienne au XVIII ème siècle." In Jeannine Monnier, Anne Lavondès, Jean-Claude Jolinon, and Pierre Elouard. *Philibert Commerson, Le Découvreur du bougainvillier*. Châtillon-sur-Chalaronne: Association Saint-Guignefort. 1993.

Leiataua AhChing, Peter. "The Polynesian History." http://asiapacificuniverse.com/asia pacific/messages8/583.html. February 5, 2004.

Lever, R. J. A W. "Whales and Whaling in the Western Pacific." *South Pacific Bulletin*. April 1964.

Lévi-Strauss, Claude. *Tristes Tropiques*. In *Oeuvres*. Editions de la Pléiade. (First published in 1955 and as *A World on the Wane*). Paris: Gallimard. 2008.

Lockwood, Victoria S. *Tahitian Transformation. Gender and Capitalist Development in a Rural Society*. Part I: "The Structural Historical Context of Contemporary Tahitian Society." London: Lynne Rienner Publications, Inc. 1993.

Lonely Planet. "Introduction to the Maldives." http://www.lonelyplanet.com/maldives

Loti, Pierre. *Tahiti. The Marriage of Loti*. Introduction by Kaori O'Connor. London and New York: KPI Limited. 1986.

Lucas, Clarize. "The Nuclear Colonialism." *Poison Fire, Sacred Earth: Testimonies, Lectures, Conclusions*. The World Uranium Hearing, Salzburg. September 17, 1992. pp. 212-213.
http://www.ratical.com/radiation/WorldUraniumHearing/index.html

Marchand, Etienne. *Voyage Round the World, Performed During the Years 1790, 1791 and 1792.* London: T.N. Longman and O'Reese. 1801. Reprinted by Cambridge, MA: Da Capo Press, 1970.

Marra John. *Journal of the Resolution's Voyage in 1772, 1773, 1774, and 1775 on Discovery to the Southern Hemisphere.* London: F. Newberry. 1775.

Mansfield, John. *Sea Life in Nelson's Time.* London: Methuen & Co. 1905.

Mateata-Allain, Kareva. "Ma'ohi Women Writers of Colonial French Polynesia: Passive Resistance toward a *Post*(-)colonial Literature." *Jouvert. A Journal of Postcolonial Studies.* (Special issue: Colonial Posts). Volume 7, issue 2 . Winter/Spring 2003.
http://www.google.ca/#9=jouvert+7.2%3A+Kareva+Mateata-Allain%2C+

------- "Orality and Ma'ohi Culture." *Shima: The International Journal of Research into Island Culture.* Vol. 3, No. 2. 2009.

Maretu. *Cannibals and Converts: Radical Change in the Cook Islands.* Translated by Marjorie Crocombe. Suva, Fiji. 1983.

Margueron, Daniel. "Tahiti ou l'atelier d'une invention littéraire. Présentation des literatures en Polynésie française." Conférence 33. Tahiti, avril 2002.
www.itereva.pf/disciplines/lettres/archi/confer33.pdf

Maude, H. E. *Slavers*, in *Colonization of the Phoenix Islands.* The Earhart Project. http://righat.org/Projects/Earthart/Documents/Maude.html.

Mauss, Marcel. *The Gift: Forms and Functions of Exchange in Archaic Societies.* (First published as *Essai sur le don,* 1922). London: Rutledge. 1994.

McKinnon, Rowan. *South Pacific.* Lonely Planet Publications Ltd. 2009.

Melville, Herman. *Typee. A Peep at Polynesian Life.* (1846). New York: The Library of the Americas. 1982a.

------- Omoo. A Narrative of Adventures in the South Seas. (1847). New York: The Library of the Americas. 1982b.

Michener, James A. *Tales of the South Pacific.* (1947). The Readers' Digest Association. 1995.

------*Hawaii.* New York: Bantam Books. 1961.

------*Return to Paradise.* New York: Fawcett Books. 1951.

Moerenhout, J. A. *Voyage aux Iles du Grand Océan.* Paris. 1837.

Montessus de Ballore, Ferdinand Bernard. *Martyrologe et Biographie de Commerson, Médecin botaniste et Naturaliste du Roi.* Paris: G. Masson, éditeur. 1889.

Moran, Michael. *Beyond the Coral Sea. Travels in the Old Empires of the South-West Pacific.* London: HarperCollins Publishers. 2003.

Napier, William. *Pacific Voyages. The Encyclopedia of Discovery and Exploration.* Garden City, NY: Doubleday and Company Inc. 1973.

New York Times. "An Indigenous Language with Unique Staying Power." *New York Times.* March 12, 2012, p. A6.

Newbury, Colin. *Tahiti Nui; Change and Survival in French Polynesia, 1767-1945.* Honolulu: University of Hawai'i Press. 1980.

Nordoff, Charles and James Norman Hall. *The Bounty Trilogy.* Boston: Little, Brown and Company. 1951.

Oliver, H. and B. R. Williams, ed. *The Oxford History of New Zealand.* Wellington. 1981.

Page, Nathan, "Guaraní: The Language and the People." 1999.
http://linguistics.byu.edu/classes/ling450ch/report/Guarani1.html

Peltzer, Louise. *Lettre à Poutaveri.* Papeete: Au Vent des Iles. 1993.

Polynesian Songs and Chants (Society Islands).
http://starling.rinet.m/kosmin/polynesia.societies.php.

Perruchot, Henri. *Gauguin. Tahiti.* Paris: Fernand Hazan. 1958.

Piddocke, S. "The Potlatch System of the Southern Kwakiutl: A New Perspective." *Southeastern Journal of Anthropology*. 21. 1965.

Poignant, Roslyn. *Mythologie Polynésienne. Polynésie. Micronésie, Mélanésie, Australie*. Paris: O.D.E.G.E. Presse. 1967.

Porter, David. *Journal of a Cruise Made to the Pacific Ocean*. Second edition. New York: Wiley & Halsted, 1822.

Rayment, Leig'h. *Alvaro de Mendaña de Nehra*. http//www.nl/tue/ñengels/discovery/mendana.heml

O'Reilly, Patrick. *Painters in Tahiti*. Paris: Nouvelles Editions Latines. Dossier 23.

Rey, Jeannot. "Le Gouvernement Hau Pakumotu passe au recrutement…" *La Dépêche de Tahiti*. April 30, 2011.

Richards, Rhys. *The Earliest Foreign Visitors and the Depopulation of Rapa Iti, 1824-30*. <http:jso.revues.org>

Robertson, George, ed. *An Account of the Discovery of Tahiti, From the Journal of George Robertson, Master of HMS* Dolphin. London: Folio Society. 1955.

Rousseau, Jean-Jacques. *A Discourse on Inequality*. (1754). Introduction by Maurice Cranston. Penguin Books. 1984.

Rowe, Kenneth H. *Where the Waves Fall. A New South Sea Islands History from First Settlement to Colonial Rule*. Honolulu: University of Hawai'i Press. 1988.

Rui a Mapuhi. *Tuvata*. Papeete: published by the author. N.d.

Rutter, Owen, ed. *The Journal of James Morrisson, Boatswain's Mate of the* Bounty. London: Golden Cockerel Press. 1935.

Sahlins, Marshall. *Islands of History*. Chicago: Chicago University Press. 1985.

Salmond, Anne. *The Trial of the Cannibal Dog. The Remarkable Story of Captain Cook's Encounters in the South Seas*. New Haven and London: Yale University Press. 2003.

--------"Voyaging with Cook." In *The Transit of Venus. How a Rare Astronomical Alignment Changed the World*. Wellington, NZ: Awa Press. 2007.

--------*Voyaging Worlds*. London: The Hakluyt Society, 2008.

--------*Aphrodite's Island: The Europeans' Discovery of Tahiti*. Berkeley: University of California Press. 2009.

-------- *Bligh: William Bligh in the South Seas*. Berkeley: University of California Press. 2011.

Samwell, David. "Some Accounts of a Voyage to South Seas in 1776-1778." In the *Journals of Captain James Cook on his Voyages of Discovery*. Edited by J. C. Beaglehole. Cambridge: Haklyut Society. 1955-1974.

Saura, Bruno. "Le placenta en Polynésie française: Un choix de santé publique confronté à des questions identitaires. *Sciences Sociales et Santé*. 22:1-23. 2000.

-------"Quand la voix devient la lettre: Les anciens manuscrits autochtones *(puta tupuna)* de Polynésie française." *Journal de la Société des Océanistes*. 126-27. Pp. 293-310. 2008.

Schmitt, Robert C. "South Seas Movies, 1913-1943." *Hawaiian Historical Review*. Vol. II, No. 11. (pp. 433-449) April 1968.

Smith, Bernard. *European Vision and the South Pacific 18?8-1850. A Study in the History of Art and Ideas*. Oxford: Oxford University Press. 1960.

Smith, Howard. "The Introduction of Venereal Diseases in Tahiti: A Re-examination." *Journal of Pacific History*. Pp. 38-47. 1975.

Sobel, Dava. *Longitude. The True Story of the Lone Genius Who Solved the Greatest Scientific Problem of His Time*. New York: Walker & Company. 1995.

Sparrman, Anders. *A Voyage Round the World with Captain James Cook in H.M. Resolution*. London: Robert Hale. 1953.

Spitz, Chantal. "Rarahu iti e autre moi-même." *Bulletin de la Société des Etudes Océaniennes* (Issue consecrated to *The Marriage of Loti*).

April-September 2000, pp. 219-226. www.lehman.cuny.edu/ile.en.ile/paroles/Spitz loti.html

------"Héritage et confrontation." Text read at the Université de la Polynésie Française, for the centenary of Paul Gauguin's death, March 2003. First published by *Ile en île*. www.lehman.cuny.edu/ile.en.ile/paroles/spitz gauguin.html

-------"Sur la Francophonie," *Littérama'ohi*, No2 (December 2002) and Ile en île. www.lehman.cuny.edu/ile.en.ile/paroles/spitz francophonie.html

------ *Island of Shattered Dreams*. (First published as *L'Ile des Rêves Ecrasés*. Papeete: Les Editions de la Plage. 1991). Translated by Jean Anderson. Wellington: Aotearoa New Zealand. 2007.

Stanley, David. *Tahiti*. Avalon Travel. Berkeley, CA: Perseus Books Group. 7th edition. 2011.

Staszak, Jean-François. "Primitivism and the Other. History of Art and Cultural Geography." *GeoJournal*. 60:40, pp. 353-364. 2004.

Stevenson, Robert Louis. "The Beach of Falesá." *Tales of the South Seas*. Edinburgh: Canongate Books Ltd. 1996.

Stewart, Frank, Kareva Mateata-Allain, Alexander Dale Mawyer, eds. *Varua Tupu: New Writing from French Polynesia*. Manao: A Pacific Journal. Honolulu: University of Hawai'i Press. 2006.

Swanky, Tom. *The True Story of Canada's War of Extermination in the Pacific*. Dragon Heart Enterprises. 2012.

Swift, Jonathan. *Gulliver's Travels*. (First published in 1726 as *Travels into Several Remote Regions of the World,* by Lemuel Gulliver). Mineola, NY: Dover Publications. 1996.

Tagupa, William. *Politics in French Polynesia 1945-1975*. Wellington: New Zealand Institute of International Affairs. 1976.

Te Ariki-tara are. "History and Traditions of Rarotonga." *Journal of the Polynesian Society*. Vol. 8. 1899. Pp. 171-78.

Thomas, Basil, ed. *Voyage of H.M.S. Pandora*. London: Francis Edwards. 1915.

Topping, Donald M. *The Pacific Islands. Part I: Polynesia*. Southeast Asia Series. Vol. XXV, No2. 1977.

Trigger, Bruce. G. *Natives and Newcomers. Canada's "Heroic Age" Reconsidered*. Kingston and Montreal: McGill-Queen's University Press. 1985.

Vaihere, Cadousteau. *Le 'Orero: le renouveau d'un antique art oratoire*. Papeete: Ile en île. 2002.

Vaite, Célestine. *Breadfruit*. New York and Boston: Back Bay Books/ Little, Brown and Company. 2006a.

-------- *Frangipani*. New York: Back Bay Books/Little, Brown and Company. 2006b.

-------- *Tiare in Bloom* New York: Back Bay Books/Little, Brown and Company. 2007.

Vercier, Bruno. "Préface" (pp. 11-38) of Pierre Loti, *Le Mariage de Loti*. Paris: GM Flammarion. 1991.

Wallis, Samuel. *An Account of a Voyage Round the World*. Edited by John Hawkesworth. London. 1773.

Warner, Oliver, ed. *An Account of the Discovery of Tahiti, from the Journal of George Robertson, Master of HS Dolphin*. London: The Folio Society. 1953.

Wayfinders. *A Pacific Odyssey*. " Polynesian History and Origin". http://pbs.org/wayfinders/polynesian. Html

Whitfield, Peter. *Mapping the World*. London: The Folio Society. 2000.

Williams, John. *A Narrative of Missionary Enterprises on the South Seas Islands*. London: J. Snow. 1837.

Withey, Lynne. *Voyages of Discovery. Captain Cook and the Exploration of the Pacific*. Berkley: University of California Press. 1987.

Worth Estes, J. "Stephen Maturin and Naval Medicine in the Age of Sail." In Dean King, with John B. Hattendorf. *A Sea of Words. A Lexicon and*

Companion to the Complete Seafaring Tales of Patrick O'Brian. Open Road Media. 2012.

Wright, Ronald. *An Illustrated Short History of Progress.* Toronto: The House of Anansi Press Inc. 2004.

FILMS AND VIDEO DOCUMENTARIES

Anthony, Na'alehu. "Papa Mau. The Wayfinder." *Pacific Heartbeat.* A Paliku Documentary Films Production. 2010.

Collingridge, Vanessa. *Captain Cook: Obsession and Betrayal.* The New World. A National Interest Program. Film Australia Ltd. 2002.

Denjean, Cécile. *Hawaï Nui Va'a. Un Marathon Mythique.* Television documentary. Grand Angle Production and Bleu Lagon Productions. 2008.

Hawaii. Hollywood film. George Roy Hill, director. Screenplay by Dalton Trumbo and Daniel Taradash, based on James A. Michener's book *Hawaii.* 1966.

Lefèvre, Xavier. "Les Marquises," *Horizons: Des Iles et des Hommes.* Television documentary. Oxala Prod. Cosmolitis Productions, Canal +, and Overseas Productions. 2011

Lenglant, Dominique. "Passion, Patrimoine: Des Pyrénées à la Polynésie. *Des Racines et des Ailes.* Television documentary. Grand Angle Production. Eclectic Production. 2010.

Miller, Timothy. *Bir Hakeim 1942. Quand la France renaît.* Cinétévé, 2012.

Moneroy-Dumaine, Benjaminne. "Polynésie: Iles de la Société." La plus grande forêt du monde. In *Horizons.* Tagra Productions avec FBD. 2012.

Mutiny on the Bounty. Hollywood film. Lewis Milestone, director. Screenplay by Charles Lederer, based on Charles Nordhoff's book *Mutiny on the Bounty.* 1962.

Pernoud, George. "L'Ile Rebelle," and "Un Cargo pour les Marquises." In Thalassa: *En Polynésie*. Grand Angle Production. 2012.

Roblin, Félicie and Matthieu Lamotte. *Nickel, le trésor des Kanak*. Zadig Productions/aaa production. 2013.

Rudolfo, Kainoa. "Ancestral Ink." *Pacific Heartbeat*. Co-production of Paliku Documentary Films, Makauila 'Oiwi TV and Pacific Islanders in Communications. 2014.

Tahiti. La Vahiné. Thalassa-Keanu TV.

Wat.tv/video/Tahiti-vahine-thalassa-john-2earf_2irat_htme

CPSIA information can be obtained at www.ICGtesting.com
Printed in the USA
LVOW04s0611160815

450291LV00016B/105/P